What Can I Do?

What Can I Do?

MY PATH FROM CLIMATE DESPAIR TO ACTION

JANE FONDA

Danelle Morton, **editorial consultant**

R A N D O M   H O U S E
L A R G E   P R I N T

Published in the United States of America by Random House Large Print in association with Penguin Press, an imprint of Penguin Random House LLC.

Cover photograph: Paul Morigi / Getty Images

Photograph Credits Pages 22, 232: courtesy of Jane Fonda; pages 114, 144, 309: courtesy of Madeline Carretero; pages 102, 119, 124, 127, 130, 132, 139, 161: courtesy of Liz Gorman; page 473: courtesy of Vanessa Vadim; other photographs courtesy of Greenpeace/Tim Aubry

The Library of Congress has established a Cataloging-in-Publication record for this title.

ISBN: 978-0-593-29479-6

www.penguinrandomhouse.com/large-print-format-books

FIRST LARGE PRINT EDITION

Printed in the United States of America

10   9   8   7   6   5   4   3   2   1

This Large Print edition published in accord with the standards of the N.A.V.H.

100% of the author's net proceeds from WHAT CAN I DO? will go to Greenpeace.

When I was young, I thought activism was a sprint, and I worked around the clock, hoping for quick change.

When I was older, I learned activism is a marathon, and I learned to pace myself.

At eighty-two, I realize it is neither sprint nor marathon; it is a relay race. The most important thing we adults can do now is join and support the next generation of climate activists ready to lead the movement.

It is to them that I dedicate this book.

Contents

What Can I Do?

Jane sits with Annie at Greenpeace USA's headquarters in Washington, D.C., at the very first Fire Drill Fridays planning meeting in September.

CHAPTER ONE

The Wake-Up Call

During Labor Day weekend in 2019, I was in Big Sur with my pals Catherine Keener and Rosanna Arquette. I have a history with Big Sur dating back to 1961, when I first ventured there by myself in search of Henry Miller. I had just read a pamphlet he wrote, **To Paint Is to Love Again**, and I wanted to meet him and talk. He wasn't there, but I ended up spending a week at the hot springs (later to become Esalen), and it was transformative.

Now here I was once again in need of transformation. I've been an environmental activist since the 1970s, installing a windmill at my ranch in 1978 and solar heating and electricity in my Santa Monica home in 1981, speaking at rallies, attending Greenpeace marches in both the United States and Canada, and later getting an electric car, stopping my use of single-use plastic, recycling, and cutting back on red meat. But I was still ill at ease. Existential angst? I had just learned that there are 2.9 billion fewer birds in North America than there were in 1970; I knew that sea turtles are strangling from tumors caused by pollution in the oceans; whales are found dead with fifty pounds of plastic in their bellies; polar bears are starving; 93 percent of children worldwide are breathing polluted air that is endangering their health; and untold numbers of people were living in the midst of oil wells and refineries that were causing them major health problems. But I hadn't really focused on what the scientists were saying.* I knew we needed to reduce fossil fuel use and invest in clean energy alternatives, fast, but these things remained a disturbing reality sitting out there somewhere, removed from me. I hadn't taken it in and metabolized it. Instead, I would wonder if perhaps humankind deserved the fate it had created. I remembered what E. O. Wilson said decades ago, which I paraphrase: God granted the gift of intelligence to the wrong species. It should have gone to non-meat-eating creatures with no thumbs such as whales and dolphins. I agreed. Just get rid of us **Homo sapiens** ASAP and things will restore themselves.

But all the while, I knew this fatalist thinking was a cop-out, and I didn't like myself for it. I'd be reminded of the recent birth of my grandson, and my two older grandkids, and the many people out there

---

* Please see Appendix A: "An Introduction to Understanding the Climate Emergency," by Annie Leonard.

fighting for a better planet. No, fatalism couldn't be for me. Yet I was compartmentalizing my grief rather than letting it into my heart.

Catherine Keener reminded me recently how, on the five-hour drive to Big Sur, she would go on an hourly rant: **What can I do? Tell me what to do! Where are the leaders? I need someone to tell me what to do!** I felt impotent, angry with myself for my inability to give her the answers she needed because I felt the same way. **What can I do?**

The very morning we left for Big Sur, I'd received an advance copy of Naomi Klein's new book, **On Fire: The (Burning) Case for a Green New Deal.** All my life, the exact book I needed without even knowing it had come to me at the perfect time and changed my trajectory. Here it was again. I began reading it the next day, and a quarter of the way through I was shaking with intensity.

Over time, I've asked myself what it was about Naomi's book that so affected me. One was the way she wrote about Greta Thunberg, the sixteen-year-old Swedish activist who, in 2018, started a movement called Fridays for Future that had inspired school strikes for climate action around the world, involving millions of students. I knew about Greta. A lot had been written about her, including that she was on the autism spectrum. But until Naomi, I hadn't understood what that had to do with the power of her connection to the climate crisis and the way she communicated about it. Naomi explained that unlike the rest of us, people with Asperger's don't look around and take cues about how to behave and feel from the people they see. They receive information pure and direct. If they study the science of climate change as Greta, a self-described science nerd, did, they aren't able to read the stark facts, feel scared for a while, and then go about business as usual. When Greta, with her unfiltered focus, her inability to compartmental-ize or cope with cognitive dissonance, read the sci-ence showing disaster was looming, she didn't believe it at first. "It can't be true, because, if it were, nobody would be speaking of anything else. All people would be doing is trying to figure out how to fix it." But when she realized the science was true and nobody was behaving as they should in a crisis, she became traumatized. She stopped speaking and eating.

Reading this was like a kick in my stomach. I knew that what Greta had seen was the truth, that, as she said, we should be behaving as if our house were on fire, as if we were in a crisis, because we are. The brave, young student was exhorting us to get out of our comfort zone and **do something**. Learning this about Greta permitted me to take the science into my own body. This was the second thing in Naomi's book that changed me: the clarity with which she conveyed what the scientists were saying in the 2018 Intergovernmental Panel on Climate Change (IPCC). Virtually unanimously, the scientists make clear that given the worsening disasters we're already seeing, and the additional warming that is already baked in because we didn't act forty years ago, we don't stand a chance at changing course in time without profound, systemic economic and social change, and they say, as of 2020, we have **a brief ten years before the tipping point is reached**. Ten years to reduce fossil fuel emissions roughly in half and then reduce to net zero by 2050 to avoid uncontrollable unraveling of the natural life-support system.

But the scientists also believe that we have the technology to make the transition in time to clean, renewable energy and that the most important factor in whether we can pull off what's needed will be collective actions taken by social movements on an unprecedented scale. **Social movements**. In my fifty years of activism, I'd been part of social movements that changed policy. For instance, in 1972 and 1973, Tom Hayden, my second husband, and I launched the Indochina Peace Campaign, and together with scores of activists we crisscrossed the country educating the people whom the then president Richard Nixon described as the Silent Majority about the Vietnam War, the Pentagon Papers, and the need to cut the funding that was shoring up the government the United States had installed in South Vietnam that was keeping the war going. Especially because the just-released Pentagon Papers revealed that a succession of administrations had known we couldn't win. The Pentagon Papers were to the Vietnam War what I think the

IPCC 2018 report was to the climate crisis: irrefutable proof of lying and deceit on the part of people in power. In the end, the aid was cut, our client government collapsed, and the Vietnam War ended. Yes, I had experienced the effectiveness of mobilizing and organizing with a clear strategy and authoritative documents to back us up.

The third thing in Naomi's book that struck me was how she explained the Green New Deal (GND). When Representative Alexandria Ocasio-Cortez and Senator Ed Markey introduced the Green New Deal resolution in 2019, I thought including issues like low-carbon jobs and environmental and economic justice in a document about climate was taking it too far and would easily be dismissed by the right as a leftist wish list. But Naomi's book made me understand justice is at the heart of solving what led to the climate crisis. The Green New Deal beckons us into a future where everyone can see a place for themselves.

When you're famous, there are so many ways to lift issues and amplify voices. God knows I've done it before to varying degrees of success. **But what can I do? What's the right way to use my platform now, when things are getting worse fast?**

Naomi's book made clear that right now is the last possible moment in history when changing course can mean saving lives and species on an unimaginable scale. A true civilizational responsibility rests on our shoulders!

I knew what I needed to do, and I felt it so strongly I was quivering all over.

"I'm going to move to Washington, D.C., for a year and camp out in front of the White House to protest climate change," I told Rosanna and Catherine over dinner at the Post Ranch Inn. "If Greta can do it, so can I."

There, I'd said it to my pals, and now I couldn't back out.

Being brave, gung ho gals, Catherine and Rosanna were all for it and pledged to join me when they could. Over the next few days we hiked, and I kept reading Naomi's book. I felt more strongly that if I got the word out, others would join me. It had happened before during the later years of the Vietnam War. At night I lay in bed trying to remember where I'd stored my sleeping bag and bivy sack that have seen me through downpours and blizzards at fourteen thousand feet. I've done a lot of camping in my life and had all the right equipment, but I'd never camped in a city. **Where will I poop and pee?** I wondered. I'm way older now and have to get up during the night more often. Rosanna, Catherine, and I studied maps of D.C. trying to pick a spot where I would set up, but I realized I didn't want a lonely vigil. What would be the point of that? I needed expert help to plan this. That's when I tried to call Annie Leonard, the director of Greenpeace USA, because they were fearless supporters of big actions and, while I wasn't certain,

I felt my action might become big. I paced in front of Rosanna's house, trying to get a phone signal, when I finally managed to reach her.

"Annie, it's Jane Fonda here, do you have a minute? I have an idea and I need your advice."

Annie assured me she did, so I rushed on. "I'm reading Naomi's book, and I've decided to move to Washington for a year and camp out in front of the White House. I want to start in three weeks. Can you help me figure it out?" See, I'm someone who, when she gets an idea, is 100 percent ready to take a leap of faith and just do it. In fact, leaps of faith are my only form of exercise these days.

There was a rather long silence on Annie's end, and then she said, "Well, Jane, that's so wonderful and I'm blown away that you're ready to put yourself out there like that, but, see, you can't camp overnight in Washington anymore. It's been made illegal after Occupy Wall Street camped there and more so in the current anti-protest climate in D.C. But let's figure out what **is** possible." She offered to set up a conference call with her, Bill McKibben, co-founder of 350.org, Naomi, the environmental lawyer Jay Halfon, and me. Time was of the essence, so Annie and I arranged to do the conference call the next day.

That day, Catherine, Rosanna, and I were visiting Esalen, the retreat center perched on twenty-seven acres of cliffs overlooking the Pacific. I figured Esalen would have decent reception for the conference call, and it felt appropriate that this possibly defining call would happen there. Generations of people seeking transformation often find themselves at Esalen, and the power it holds to create change is very much about its topography. It's edgy. The sea off Big Sur is where the Arctic current meets the warm Pacific current. Land and sea touch each other at the base of jagged cliffs. It's all the sharp edges and fierce winds, I think, that encourage people to break with their pasts, to look beyond dogma, and to examine new ways to be effective in the world. I remembered arriving there, a young woman in my midtwenties, the same age as many of the activists now driving the movement to stop climate change. Big Sur's wild nature, those cliffs, was an important part of my transformation from child of the buttoned-up 1950s to someone who wanted to shake off the "good girl" shackles of my youth. In that way, it seemed right that I found myself in Esalen at this moment when I was rethinking how I could work to serve something larger than myself.

Turns out there was very poor cell phone service, but there was a bright red old-fashioned outdoor phone booth and a place where I could gather an hour's worth of quarters.

Bill McKibben suggested that because camping was no longer permitted, perhaps a once-a-week pro-test that involved civil disobedience would be a better idea. Fridays had been claimed by Greta Thunberg and the student climate strikers, but the youth had also called on adults to join them. "Maybe you could do an action on Fridays as well." Bill referenced what Randall Robinson, executive director of TransAfrica, had done in the mid-1980s, marching in front of the South African embassy in D.C. every day, committing civil disobedience by sitting down in the middle of Massachusetts Avenue, calling for freeing Nelson Mandela and ending apartheid in South Africa. "It was a very successful protest," Bill said. In the beginning there were ten to twenty people, which grew to hundreds, and it spread nationwide until, in 1986, the first antiapartheid bill was passed in Congress.

Annie agreed, noting how important civil disobedience on behalf of climate has become. "For forty years, we've been polite, we've shared the science, we've petitioned, we've marched, rallied, written, pleaded. We've used all the levers of democracy available to us, and our elected representatives haven't listened. Now we have to do more, step it up. Risk arrest if that is what it takes. It's the fossil fuel industry who've led us to this. It's time to be bold. Science demands it. Morality demands it. The moment demands it."

Yes!

I dropped in more quarters and tried to breathe, not just because it was getting really hot in the phone booth, but because I could feel in my body that this was right. I was ready for this. I'd been getting ready for this my whole adult life . . . a weekly action that culminated in nonviolent civil disobedience. And I wouldn't have to worry about pooping.

I wanted to start in a few weeks, so we decided I should go to D.C. as soon as possible to work out details and meet with some key environmental groups, including the student climate strikers, to get their input and buy-in. I began planning for the maximum time I could spend in D.C. before I had to get ready to film our seventh and last season of **Grace and Frankie** on January 27, 2020. It added up to four months, fourteen Fridays.

I started a list of the bare essentials I'd have to bring. I asked Debi Karolewski, who began assisting me in the early 1980s, to come with me. I knew I'd need my little dog Tulea, a fifteen-year-old Coton de Tulear. I couldn't imagine being without her for four months. I also knew how sad I'd be to not see my two-and-a-half-month-old grandson, Leon, for that long.

I went about canceling everything on my schedule. Fortunately, I hadn't booked any acting jobs during that time, but that also meant I had no source of income except a number of speaking gigs I had contracted for, and I soon realized they could sue me if I canceled them. Also, five of them were in and around Los Angeles with Lily Tomlin, and that would mean taking airplanes to and from Washington. Here I am, trying to reduce my carbon footprint, speaking out against fossil fuels, yet I'd be flying.

I spoke with Annie and several others about this contradiction. We weighed not flying against the possible good I could do in the climate movement by carrying out the Friday actions, and we concluded that the actions were more important. In the process of discussing this, I came to realize that as important as our individual lifestyle decisions are, they cannot be brought to scale in time to get us to where we need to be by 2030. I realized the importance of progress and not perfection. It's structural change, new policies, that we need to focus on while at the same time continuing our individual commitments to the planet. Maybe the Friday actions would help bring about that policy change.

On September 27, en route to the Los Angeles airport with Debi, my little Tulea had a seizure. Tulea had already been diagnosed with an age-related enlarged heart and damaged heart valve. My own heart sank as I came to grips with the fact that I'd have to leave her behind. I conjured up the image of Greta. You have to leave your comfort zone. I left. It wasn't easy.

Over ten days, the essential core team came together with Annie's guidance. She brought in DC Action Lab, which handles logistics for all the big actions in D.C. Samantha "Sam" Miller with that organization has probably trained ten thousand people in civil disobedience and getting arrested.

In turn, Sam brought in our digital team, including Vy Vu, a young Vietnamese artist and student, to do our weekly topic-focused posters. Vy had to design the posters at night, after work, and at breakneck speed. When I first spoke to her by phone, I asked where she came from in Vietnam. "Hanoi," she replied. "Oh," I said, "I've been there a few times." And she asked, "Really? How come you went there?" I loved it. There was no reason Vy should know all I had done in the 1970s to oppose the Vietnam War, decades before she was born, or for her to know that because of my trip to Vietnam in 1972, in some political circles they still refer to me as Hanoi Jane.

With just fourteen days to go, my team still didn't have a name for the action. We had assembled on an emergency basis to come up with one, but after more than an hour of throwing out different ideas, we finally gave up and decided to reconvene the next day. As we were packing up to go home, Greg, the sound-man from the documentary film company that had been shooting our work, capturing the process, took off his headphones and said, "What about Fire Drill Fridays?" We all looked at each other and burst out laughing. That was it!

As with all subsequent ones, the first big meeting in D.C. was convened in the Greenpeace office, and about a dozen people were there representing the Sunrise Movement, Friends of the Earth, Climate Action Network, Hip Hop Caucus, Oil Change International, and, of course, Greenpeace. It was critical to get broad movement participation and buy-in. I laid out my vision.

We spent a long time discussing who we were targeting with these actions. Climate deniers? Conservatives? Independents? Activists? And we decided we needed to aim for people who acknowledge there's a man-made crisis; who support the climate movement and are thinking about maybe doing more but don't know what that could be; who are confused, paralyzed, or tilting toward hedonism or fatalism. Hedonism being the thinking that because everything's going to hell anyway, I might as well eat, drink, and tune out with shopping or debauchery. Fatalism being the thinking that I had started to tilt toward prior to Labor Day: Humans have done so much harm we don't deserve to survive.

The bulk of the meeting was spent defining our demands and calls to action. We whittled it down to the three essential demands without which we will never meet the Paris Agreement on climate goals in a sustainable, fair way.

SUPPORT THE GREEN NEW DEAL

NO NEW FOSSIL FUEL EXTRACTION

PHASE OUT EXISTING FOSSIL FUELS WITH A JUST TRANSITION TO CLEAN RENEWABLE ENERGY

We decided our calls to action would be the following:

VOTE:

Vote for the climate in every election up and down the ballot. Vote for candidates who are in favor of a Green New Deal and a bold and responsible transition from fossil fuels to clean renewables.

VOICE:

Make your voice heard. Initiate conversations about climate with your family and colleagues. Tell your candidates or elected officials that climate can't wait. Call them, sign petitions, and go to their town halls. Write letters to the editor of your local paper. Divest from fossil fuel companies and invest in a sustainable future.

USE YOUR FEET—WITH OTHERS!

Join an organization working for real climate solutions. We are stronger together than we are alone! Join marches, do outreach, recruit friends to join. Show up for the communities on the front lines of the fossil fuel economy. Show up for ALL OF US.

Listen to communities most impacted by climate change, and if you can, put your body on the line wherever people are fighting on the front lines. Start by joining a student climate strike or Fire Drill Friday action in your own community!

After reading Naomi's book, I recognized that there was so much more that I wanted to know and wanted to bring to a bigger audience. People needed to know what was really happening! I suggested weekly teach-ins before each rally, each one focusing on a different aspect of the climate crisis and featuring experts, scientists, and activists. Karen Nussbaum said rallies weren't ideal places for teach-ins but we could hold them the evening before, maybe in a theater in D.C., and film them.

Karen is a labor movement activist who has been my friend since the early 1970s, when she was an organizer with Tom and me in the Indochina Peace Campaign. She is the founder of Working America, the community-outreach arm of the AFL-CIO, on whose board I sit, and in the 1980s she founded 9to5: National Association of Working Women. It was she who inspired me to make the film **9 to 5**. Karen was here to help us find ways to bring more of the labor movement into collaboration with the climate movement; in addition, her husband, Ira Arlook, had agreed to serve as the press coordinator for our weekly actions.

That was when Carla Aronsohn on our digital team asked, "Well, why not do the teach-ins digitally. We're setting up your website, and we can do them livestreamed and archive them on your site." And that is how our Thursday evening teach-ins came to be.

In the days following, I had an important meeting with eight of the leading D.C. student climate strikers, ranging from fourteen to twenty-three years old, from the Sunrise Movement, U.S. Climate Strikes, Zero Hour, and Fridays for Future. Yes, the youth in these organizations had called upon adults to step up and join the fight for their future, but an aging movie star bopping in from Hollywood whose action would likely get a lot of attention on their Friday? I needed and wanted their blessings.

It was a real learning experience. I was amazed at the students' depth of passion, organizing smarts, and sensitivity to the need to center vulnerable communities and indigenous peoples. Even the youngest wasn't afraid to correct me if she felt I was on the wrong path. It soon became clear that they welcomed this addition to their school strikes, but there were many things we had to iron out with them before my first action, which was only ten days away. Would students stand with me for that action? Who should it be? Sebastian Medina-Tayac and his seventeen-year-old sister, Jansikwe, are members of the Piscataway Indian

Nation on whose land we would hold our actions at the Capitol. It was decided Jansi would open by welcoming us to her people's land. Jerome Foster II, a seventeen-year-old African American student and founder and director of OneMillionOfUs, which is mobilizing a new generation of young people to register to vote, has been climate striking every Friday for a year in front of the White House. He would march from the White House to join us at 11:00 a.m. and speak.

Things were coming together. It was no longer one person's idea. It was taking shape and evolving into a team effort, and other organizations were feeling included and heard. In fact, it felt meant to be.

Support the
green new deal

No new fossil fuel
extraction

Phase out existing
fossil fuels with a just
transition to clean
renewable energy

Jane speaks at the first Fire Drill Friday, holding up a copy of Naomi Klein's book On Fire: The (Burning) Case for a Green New Deal.

CHAPTER TWO

The Launch

It was a beautiful day. On the morning of October 11, 2019, we all gathered in the United Methodist building near the Capitol for a pre-rally briefing. I wore the red coat I had bought on sale a few days before at Neiman's and a black-and-white-checked cap to hide the two inches of gray hair I was letting grow in. Hair epiphanies have always accompanied my life transformations, and going gray felt right for my new (and maybe final) turning point. Little did I know that the red coat would become a popular Halloween costume a few weeks later and a pop culture totem.

I was scared and I hadn't slept. It's that fear: **What if I give a party and no one comes?** We thought it would work, but none of us were certain if this weekly action would actually gain traction and make a difference.

The core team was all there plus two of the speakers

Jane stands with Naomi Klein, at the morning briefing ahead of the launch of Fire Drill Friday on October 11.

and about a dozen activists who wanted to engage in civil disobedience with me and risk arrest, like Karen Nussbaum, Annie Leonard, and Steve Kretzmann, director of Oil Change International. My step-granddaughter, Vasser Turner Seydel, showed up in a bright red sweater. I had filmed her birth and now here she was, twenty-four years old and ready to risk arrest with Grandma Jane. I was moved and impressed.

I greeted people as they came into the somewhat cramped room filled with a long, oak refectory table running down the middle and sturdy chairs that reminded me of the furniture at Emma Willard, my old boarding school.

At 9:30, Sam, our head of logistics, quieted us all down and began the briefing that she would do for every subsequent Fire Drill, explaining where we would go when we left the building, who would speak when, where we'd commit the civil disobedience, and what to expect if we were arrested. At her direction, everyone planning to risk arrest took off their jewelry and made sure we had $50 and an up-to-date photo ID on us. If they didn't have the $50, we gave it to them. I saw Dr. Sandra Steingraber, a distinguished scholar in residence at Ithaca College and a biologist who would be speaking at the rally, taking off her necklace. **Ah, no ordinary scholar, she's risking arrest! How cool!**

Then Firas Nasr, from our digital team, told us that because we'd be putting our bodies on the line, we'd best get into our bodies, and he took us through a brief meditation, bringing us inward, which I needed to calm my nerves. That was followed by three from-the-gut martial arts shouts to ground us. It worked.

Flanked by the speakers and friends holding the Fire Drill Friday signs with our three demands, I stepped out into the warm, welcoming day. I felt tears running down my cheeks. Here we go.

We were met by a phalanx of television cameras and photographers, walking backward as they filmed us marching and chanting. There were more of them than there were of us. The interviews Ira had arranged for me to do with big media outlets had clearly generated attention.

As we crossed the intersection between the Supreme Court and the Capitol, their majestic columns and curves spoke to me of history and of moral authority. Would that the goings-on inside right then matched these imposing facades. The impressive buildings added to the feeling that our little group was ragtag and inconsequential. Our chants lacked confidence, and there weren't enough of us to make a lot of noise anyway. At the stage where our banner was erected, I could see our supporters. Sparse. Fifty at most. Well, it was a start, and hopefully the media coverage and our livestream of the rally would expand our reach.

I welcomed people and thanked them for coming. "I'd like all of you to think about this," I said. Good. My voice felt strong. "The same toxic ideology that took this land from people who already lived here, that kidnapped people from Africa, turning them into slaves to work that stolen land, and justified it by saying that those kidnapped and displaced people were not human beings, cut down the forests, and exhausted the natural world just as it did the people.

The size of the crowd at the launch.

This foundational, extractive ideology of commodification is the same one that has brought us the human-driven climate change that we're facing today." And with that, I invited Jansikwe Medina-Tayac to come up and welcome us to her people's land.

Jansikwe Medina-Tayac kicks things off at the first Fire Drill Friday.

Jansi, with her long curly hair and blue jeans, stepped from behind the banner to the mic. I was glad that we had already spent time together. Without the young climate activists who had been striking every Friday supporting us, these actions would not have worked out. Giving them this platform was proof of our unity.

"Hello, everyone, my name is Jansikwe. I am seventeen years old, and I am a member of the Piscataway Indian Nation." Her voice was strong and confident. "Our traditional territory spans all the way from the Potomac River down to the Chesapeake Bay. I'm here not only to welcome you to this space, but to remind you that indigenous people have been fighting for this earth since the early 1600s. We are the original protectors of this earth and its indigenous ways of life."

As I listened to this young indigenous woman, I marveled at how remarkable it is that despite all that European colonizers have done to the original inhabitants of this land, many are still willing to offer welcome, advice, and guidance about how we must live in relationship with nature and each other. In the course of the fourteen Fire Drill Fridays, I would learn so much more about the critical role indigenous peoples play in fighting climate change and how disproportionately impacted they are by fossil fuel extraction.

"We need to recognize and change the system that encourages us to take and take and never give back," said Jansikwe. "Native kids in places like Standing

Rock should not have to take time off from school to fight for their lives. We need every person, every helping hand, every heart in the world, to come together and help end this attack on our planet." She stepped off the platform to rejoin her mother and brother.

Because this was the launch, I wanted to take this opportunity to explain why I had moved to D.C. to hold these actions.

"So much is happening in the news. There's so much noise, right? We have to ensure that the climate crisis remains front and center, and that's why we're here." I spoke about how Naomi Klein's book, Greta Thunberg, and the student climate strikers had inspired me to get out of my comfort zone. "So the question for the rest of us is, what are we willing to give up? What time and energy will we devote to it? What sacrifices will we make?

"I'm standing with the young people. I want to help lift their message. Greta Thunberg said we have to behave like we're in a crisis. Our house is on fire. And so we're calling these rallies Fire Drill Fridays.

"Every Friday for the next four months at 11:00 a.m., we're going to be right here. And every Friday, we're going to focus on a different aspect of the climate crisis."

I listed our demands: Pass a Green New Deal, stop fossil fuel expansion immediately, phase out fossil fuels as soon as possible but definitely within thirty years, secure a fair deal for workers and communities most impacted by this transition. Our other mission was education. Each Friday we would invite scientists, experts, and people from frontline communities along with celebrities to focus on a different aspect of the climate crisis. "We need to understand that at least 97 percent of the world's climate scientists agree that we are facing a drastic emergency, that it is man-made, that we have a little over a decade before the tipping point is reached. Ten years to reduce fossil fuel emissions roughly in half and then reduce to net zero by 2050. There aren't two sides to this story."

When I was done, I brought up seventeen-year-old Jerome Foster, who startled me with his command of the stage. Clearly, he was accustomed to speaking at rallies. I had spoken at many myself over my years of activism, but not on a regular basis, and of late my throat had had a tendency to close up at rallies, making it difficult for me to project. I knew I had to be careful, warm up my voice, and not overdo the chants.

Jerome focused on the need for unity. "Change can only happen with unity. Change can only happen when everyone is at the table, because only in the cracks of division can corruption seep in, and only in the cracks of division can pollution seep in. We are a part of one earth, one pale-blue dot that's in the middle of an endless sea of blackness. What we're saying is that we must act as such, we must act as if we are one people, one planet, and one globally interconnected nation." Then he exhorted people to vote and

Jerome Foster at the first Fire Drill Friday.

to write to their elected officials, and he led us in a chant:

> **Take it to the polls!**
> **Take it to the streets!**
> **Take it to the polls!**
> **Take it to the streets!**

Next, Sandra Steingraber, the biologist and scholar, took the mic. She described how fracking "swung a wrecking ball at our climate system" and how it exhumed the life buried deep in the earth thousands of years ago to quell our unquenchable thirst for oil.

"Diatoms and sea lilies and squid. We blow up a

cemetery of prehistoric sea creatures that we rename fossil fuels in order to light their bodies on fire in the crematoria we call power plants and internal combustion engines," Professor Steingraber said.

A brilliant mind, Professor Steingraber illuminated connections to bring a fresh focus on the destruction inherent in the way we live. It was hard to hear her list all that we had lost and what we'd be losing in the near future. Fracking destroys our drinking water and discharges dozens of carcinogens into our atmosphere. The carbon released by the oil it collects destabilizes our oceans, our food supply, and our water supply, sending millions of climate migrants in search of safety.

"We are losing the world's fish stocks and coral reefs," she said. "We are losing insects, pollinators, and reliable rainfall. Failed harvests are driving migration crises across the globe. Does that sound like a smart energy system to you? No, it's primitive and crude. Crude as in crude oil."

The other crude effect of fracking goes mostly unnoticed, she pointed out. Energy companies sell off the by-products of fracking to chemical companies to make the single-use plastics that are choking our ocean life and often end up burned in incinerators—adding to climate change and pollution.

The passion I heard in her voice came from personal experience. Forty years ago, when she was just twenty, she was diagnosed with a rare cancer that her doctor said was likely caused by environmental carcinogens.

Immediately she decided to become a public health research scientist instead of a doctor, but at that moment she wasn't sure she'd live long enough to realize any of her dreams.

"Like any teenager, I had felt immortal. After the diagnosis, the future was uncertain and, if it existed at all, was full of dread and suffering," she said. "So my message to youth today is, I get it. When you say that your future has been stolen and held hostage by the actions of others, I understand. When you see grown-ups all around you carrying on as if everything is still fine and there is no catastrophe, I know that unbearable feeling, too. We've all become cancer patients now."

Dr. Steingraber's speech was a revelation: The bodies of marine animals that died 400 million years ago are being weaponized to destroy the bodies of sea creatures living in our oceans now.

I hadn't made that connection. Nor the fact that making plastics was a way for oil companies to use the chemical industry to deal with their waste disposal, thereby creating the horrendous waste problem the rest of the world is trying to solve. What an interesting woman, Sandra Steingraber. A biologist, an activist with

Sandra Steingraber speaks.

the creative instincts of a poet who, despite the dangers she described, left us with a message of hope, the same one her adoptive mom had given her when she started her battle with cancer: "Don't let them bury you until you're dead."

"Friends, I am here today with science in my hands, with love in my heart for the whole sunlit planet and all of those who walk its surface, and with a cancer survivor's fierce determination to fight for life, no matter what the odds," she said. "The fossil fuel industry will not bury us. We will live to bury them."

I found her speech so profound, so striking. The imagery she used to stir the crowd to action stayed with me a long time after she finished. As soon as she left the podium, I asked her for a copy of her remarkable speech, because I knew I'd read it again.

When the rally was over at noon, we marched behind the big "Fire Drill Friday" banner, chanting,

Tell me what democracy looks like!
This is what democracy looks like!
The seas are rising and so are we!

We marched past a line of police cars toward the Capitol steps, the press in front of us walking backward, occasionally tripping and running into things. The line of about ten police officers standing on the steps began to move to the side to make room for us. They were well practiced. There were around sixteen of us who mounted the steps and turned around to

The group risks arrest at the launch.

face the small crowd and media people, which was being pushed back by another line of police. "Move back, people. Move all the way back."

And soon there was a wide distance between those of us risking arrest and the others, though we all kept chanting. My step-granddaughter, Vasser, was right next to me along with Carroll Muffett, director of the Center for International Environmental Law; Steve Kretzmann, director of Oil Change International; Wendy Fields, director of Democracy Initiative; Medea Benjamin with CODEPINK; the twenty-four-year-old climate activist Sebastian Medina-Tayac; Annie Leonard, director of Greenpeace USA; the Greenpeace staffer Madeline Carretero; and Sandra, the poetic biologist.

Annie Leonard and Maddy Carretero get arrested on the Capitol steps.

The head police officer gave us the first warning that we must leave or risk arrest. We kept chanting. Then a second warning. A few people left, but we kept chanting. The final warning, and that was it. One at a time, officers secured our hands behind our backs with white plastic zip ties. They hurt.

I wasn't scared. I had been arrested before, but this was my first arrest for civil disobedience. In 1970, I had been arrested in the Cleveland airport returning from Canada, where I had just begun a national speaking tour about the atrocities of the Vietnam War.

Back then, the arresting officer told me the Nixon White House had ordered my arrest. The police took my notebooks and my address book and dozens of little plastic bags containing the vitamins I took with each meal. The charge was drug smuggling. I was put in a cell with a woman in the throes of drug withdrawal. I didn't know what was going to happen to me.

This time was different. I stood chanting on the steps of the Capitol, energized. I was doing what I had wanted: putting my body on the line and aligning

Jane Fonda gets arrested. Behind her stand Annie Leonard and Maddy Carretero.

myself fully, body and spirit, with my values. I felt empowered. The people around me seemed to feel the same. As each of us was taken by an officer to the waiting vans, people cheered, clapped, and chanted in support. It felt good.

When I arrived, several women were already sitting in a row on one side in the back of the van, which was divided in two. Even though I'm strong, it was a high step up with nothing to hold on to and my hands cuffed behind my back. The officer helped by boosting me up by my behind, and I flashed onto a scene in the sixth season of **Grace and Frankie** when Peter Gallagher had to help me into an SUV the same way because Grace is too old to do it on her own. Oh, well, Fonda. This is your new reality.

It took an inordinate amount of time before we finally arrived at the police station, where we were off-loaded and taken inside. There, we were searched, and everything we weren't wearing, including glasses that weren't prescription, identification, and money, was put in a clear plastic bag and marked with our name. Given what we were committing civil disobedience for, the extensive use of plastic was glaring. My red coat was so new I hadn't yet realized it had real pockets that had to be unstitched open. Hence, my money and driver's license were in my bra, which the officers seemed to find amusing. They had to release me from the cuffs so I could get them.

The thin white plastic handcuffs were cut off, tossed,

and replaced by thicker black ones that screwed into place and were apparently recycled after use. Then we were led into another room containing two cells, each painted hospital green, each with a sleek metal slab of a bed/bench and an all-metal toilet. There was nothing that could be removed or broken off. No sharp corners.

In the cell with me were Vasser, Sandra, Medea, and Wendy Fields. I knew that Vasser's father was nervous about her risking arrest, and I was happy and proud that she felt empowered by her decision to do it. Karen, Annie, Maddy, and the other women were in the next cell over.

While I did wall squats (hey, you gotta seize the opportunity when you can), we talked about climate's connection to many things, from democracy to war to health, and I got more ideas for upcoming teach-ins.

Turned out that both Medea and Sandra had known my ex the late Tom Hayden. Sandra told me that back in 1965, students at the University of Michigan had been protesting the escalation of America's involvement in the Vietnam War and of military research being done on campus. Tom, who was the editor of **The Michigan Daily** at the time, and three thousand other protesters took over Angell Hall, a kind of hallowed place on campus. All night long they held seminars and gave speeches about the history and culture of Vietnam and about Lyndon Johnson's bombing campaign. And right then and there, it was labeled a teach-in, a new concept. Instead of a walkout or a sit-in, this was a teach-in. Soon the idea of teach-ins spread to other schools.

"In 2015, fifty years after that original teach-in," Sandra recounted, "Tom and I were invited back to a teach-in on climate change. It was held in the same hall as the original teach-in, Angell Hall. I believe that was Tom's last big speech before he died."

This story brought up so many emotions for me. I thought about the fact that next Thursday we would have our first teach-in. I was happy for that link to Tom. I missed Tom and wished he could be with us now to guide and advise us. Strategy was always Tom's strength. I wondered what he'd think about me being in this jail cell with Sandra and Medea following civil disobedience. I know he would have loved the fact that my step-granddaughter was in there with me. He always favored cross-generational organizing.

To be truthful, though, I had been privately wondering if I could have gotten up the nerve to move to D.C. and launch the Fire Drill Fridays were Tom still alive. I had always been in awe of his fierce intelligence and broad movement-building experience, always feeling myself the student to his teacher. Had he been the least bit skeptical of my idea, which he often was for reasons that sometimes felt arbitrary and personal, I might well have backed off. But then again . . . maybe not. I was also aware of how much more confident I had grown in the thirty years since Tom and I divorced. Maybe I would have been capable of laughing it off as part of Tom's quirky ego that could no longer intimidate me.

After three hours or so, people started to be processed out, but I was unable to see who went first. An officer came to talk to Medea about the fact that this was her third arrest of the year and that they were considering keeping her overnight. Instead, they gave her a court date to appear before a judge and let her go, telling her that if she had another arrest before her court date, she would definitely spend a night or two in jail. I was learning the rules that would come in handy in the coming months.

I was one of the last to be released, and as I sat in the cell alone, I had time to reflect on this action that we had hurriedly pulled together based on my sense of urgency the month before. I felt proud that we had done this. I loved the voices we included in this launch: respect for the native land, for youth, and for science. A good beginning but not nearly enough. There was so much more I wanted to learn about all the different topics we planned on covering. Where would we be a month from now? Two months from now? I was ready for that!

About four hours after we were brought into the cell, I was thumb printed, paid my $50 fine, and got my possessions back. Out I went into the glaring sunshine, where I saw the team, plus Debi, my assistant, Annie, Maddy, and Karen, all clapping and cheering and offering chips, tangerines, and water. I was surprised to see that people had waited, but Sam explained that this is what's called jail support. I hadn't expected it, all those people there to hug me and thank me for being willing to get arrested for the cause made it different and special. And for every one of our fourteen Fridays that involved arrests, rain or shine or frigid weather, jail support was always there, and in time, when I could no longer risk arrest, I would be part of it myself.

That day cemented for me how important it was to work across movements, to bring together people who work on democracy and women's issues, indigenous issues, antiracism, peace, labor, and more. It's not just a matter of good movement manners. We didn't need to do that; we could have just started this on our own. But taking the time to engage people across movements makes us stronger. There is a saying: "If you want to go fast, go alone; if you want to go far, go together." The more I was learning about the climate crisis, the more I knew that building a community was how we would grow the army that was needed to change the way this country does business—literally, and for the long haul.

As I thanked everyone for sticking around and started to leave, a Fox TV reporter showed up and stuck a mic in my face while his camera rolled.

"Why have you done this, gone to jail?" he asked with an edge in his voice.

"To get you to cover climate," I replied, and got into a waiting car.

Sam Waterston gets arrested for the first time.

CHAPTER THREE

The Green New Deal

I was bleary-eyed the morning after our launch, but when I looked at my iPhone, I realized Fire Drill Friday had gone viral. Ira reported later that there were 8,670 press articles online, on TV, and in print with amazing images of the entire thing, especially my arrest in my red coat. People had been calling and texting me from all over to congratulate me and thank me for standing up for the planet. I thought, **Oh my God, what's happening?**

I did twelve more interviews that week, and we had another four-thousand-plus stories in the media . . . all around the world. And while all this was going on, I had to study up on the Green New Deal, the focus of the next week's first teach-in on Thursday and Fire Drill Friday, when I'd be joined by my friend the actor Sam Waterston and two high-profile experts on the Green New Deal. At my age, the whole livestreaming thing was foreign, and I was nervous.

Wednesday morning, one of the experts let me know that his young daughter was ill and he wouldn't be able to come, and Wednesday afternoon the other one told Greenpeace that she was too ill to attend. Clearly Sam and I couldn't hold a teach-in on the Green New Deal by ourselves. I went into meditation mode to quiet my growing panic. Was it always going to be this dicey?

Fortunately, Greenpeace got a Thursday morning commitment from a young activist, Joanna Zhu, a fellow within the Sunrise Movement's political department who, though not a full-on Green New Deal expert, knew enough to explain it to our teach-in audience.

The Teach-In

I arrived at the Greenpeace conference room, which we were using as the digital teach-in set, early Thursday evening. Firas had suggested that the teach-ins have a homey vibe, so I had rented furniture online and created a comfortable sitting area with couch, chairs, and coffee table in a corner of the conference room. I made sure Naomi's book was prominently displayed on the coffee table, pilfered some plants from staff desks to dress up the side tables, and arranged the Fire Drill Friday and Green New Deal posters appropriately. It was the first time I'd seen everything all put together, and I must say the effect worked pretty well.

The documentary crew was there to film, and Carla, with the digital team, set up an iPhone on a tripod to livestream the whole thing.

Because there was no teleprompter, I had note cards for the opening and closing, as I would for every subsequent teach-in, with questions for the guests and additional points I wanted to make if there was time.

At 7:00 p.m. sharp, Carla counted down, 3, 2, 1, and off we went.

"The Green New Deal is not a policy or even a bundle of policies," I said. Talking to the tiny camera on a tripod about six feet away felt funny. It was hard to believe that that little camera could potentially be the eyes and ears of tens of thousands of interested people from around the world, which, as it turned out, it would become.

I was so jacked up that I introduced Sam Waterston as Sol Bergstein, which is the name of his character in **Grace and Frankie**, and was momentarily confused when Sam kept saying, "No, Jane. I'm not Sol Bergstein."

"The Green New Deal is a framework," I said, "an integrated, systemic response to the multiple crises of climate, of democracy, and of equality."

When I had finished my opening, I asked Joanna to explain the Green New Deal to us. She spoke for a minute instead of five and then turned to me. My heart sank. **OMG, that's it?** I was going to have to wing it.

The three of us did a pretty decent job describing

what the Green New Deal is, but let me just summarize it here so we're all on the same page.

THE GREEN NEW DEAL

The Green New Deal details the steps our country needs to take to prevent the most severe impacts of a changing climate. Introduced by Representative Alexandria Ocasio-Cortez and Senator Ed Markey in February 2019, the Green New Deal warns that for us to survive and to avoid huge climate catastrophes, we must prevent global temperatures from rising 1.5 degrees Celsius (2.7 degrees Fahrenheit) above preindustrial levels. This requires a 40 to 60 percent reduction in global carbon emissions by 2030—in a decade—and then further reduction to net zero by 2050.

As the country that has produced 20 percent of these emissions through 2014, the United States must lead on this issue. We have the capital and the technology to show the world how to transform our way of life to one that is healthy and sustainable for all. The Green New Deal describes a ten-year mobilization to save the planet, where the government funds thousands of good-paying jobs that will bring clean water and clean air and access to nature and to healthy food for all of our citizens while establishing renewable, zero-emission energy sources to power our economy.

This is why having "New Deal" in the title makes sense. To tackle such large issues that affect so many people, the effort is similar to the New Deal and World War II, dire times when the federal government rose to the challenge. We need just that kind of effort to head off climate catastrophe. And the Green New Deal shows us the way.

Much like the New Deal, the Green New Deal begins with jobs, good ones: retrofitting and upgrading buildings for maximum energy efficiency; updating water quality and conservation; remediating toxic waste dumps; and generally increasing efficiency, safety, affordability, comfort, and durability. As with the jobs created under the New Deal in the 1930s, people would be trained in skills that would be in demand in the new economy created by the transition off fossil fuels. The work to be done is local, and the local communities will be the ones that benefit from these changes. While polluting industries are phased out, other sustainable industries will replace them, spurring economic development.

The changes wouldn't just come in the cities either. The Green New Deal would also work with farmers and ranchers to remove pollution and greenhouse gas emissions from the agricultural sector by investing in electric cars and trucks, farm machinery, buses, and high-speed rail. The GND supports family farming, invests in sustainable land use practices, and restores the natural ecosystems through proven low-tech solutions that increase soil carbon storage and clean up hazardous waste sites.

The genius of the Green New Deal is its inclusiveness. It doesn't just focus on greenhouse gas emissions. Vulnerable communities including indigenous peoples, communities of color, poor and low-income workers, and people who live in rural communities are the primary victims of pollution and extraction of natural resources; under the Green New Deal they would be the beneficiaries. The GND jobs program would focus on opportunities for workers affected by the transition off fossil fuels and training and advancing the right of all workers to organize, unionize, and collectively bargain would be guaranteed, and the overseas transfer of jobs and pollution would be stopped.

Truthfully, I was originally upset that the resolution seemed to go far beyond climate. But the more I learned, the more I realized it couldn't be any other way. We can't just move to green energy and leave it at that. The transition to a fossil-free world must be framed in such a way that people and communities that have been most directly harmed by the old energy system can be made whole and given agency over their energy delivery systems. Had there been a Green New Deal in place in early January 2020, COVID-19 would have been much easier to contain and our health-care system much more resilient.

With Fire Drill Fridays, I wanted people to see the climate crisis as a chance to create a world of new possibilities for fairness, prosperity, and good health. If enough people demand it, just as they did with the original New Deal, we can make it happen. This will mean the climate movement has to stretch in a few ways.

First, the conventional nature/wildlife conservation people need to widen their perspective on "environment." Environments where fossil fuels—coal, oil, and gas—are drilled and processed may be industrialized and polluted, but they are environments nonetheless. Many people in the United States never see these communities on the front line of the fossil fuel economy. Too often they are out of sight and out of mind. Through the Fire Drill Fridays platform we were creating, I wanted to use my celebrity to make the real impacts of fossil fuels visible to everyone. To pass a Green New Deal, the environmental movement must unite and grow to include people who hadn't considered their environment something they had the power to change.

Second, the climate movement needs to approach fossil fuels from two sides: using less and making less. See, when it comes to efforts to reduce fossil fuel emissions—which are the primary driver of the climate crisis—environmentalists and elected leaders have historically focused on the **demand** side, meaning they support approaches to reduce the **demand** for fossil fuels with new green sustainable energy, electric vehicles, and energy conservation. Those are all great, and we need them all, but to add up to enough emissions reductions to keep below that crucial goal of 1.5 degrees Celsius, we also need to reduce the **supply** of fossil fuels being extracted, pumped, and fracked. Without a reduction in fossil fuels at both ends, it is easy for people to look around at the growing number of solar panels and electric vehicles and think the problem is being solved while being unaware of what's happening to people different from them in environments they may never be exposed to. To pass a Green New Deal, the traditional environmental movement must grow to include issues beyond conventional conservation and wildlife concerns and include people who might be very different from themselves. That's how we will build a movement big enough to win and ensure we leave no one behind.

Critics say shifting away from fossil fuels will wreck the economy, but many experts see enormous potential to create jobs and stimulate the economy through a Green New Deal. The Stanford University professor Mark Jacobson, in his study published in the journal **One Earth**, described how going 100 percent green could pay for itself in seven years. Revamping power grids and remaking transportation, manufacturing, and other systems to run on wind, solar, and hydro power would cost $73 trillion. But, Jacobson says, that would be offset by annual savings of almost $11 trillion. Jacobson claims in his paper that studies among at least eleven independent research groups have found that transitioning to 100 percent renewable energy in one or all energy sectors, while keeping the electricity and/or heat grids stable at a reasonable cost, is possible.

During our livestream, I asked Joanna what she would say to the critics who call the Green New Deal a "wish list for the far left"? I liked her response: "It isn't just a wish list; it's a to-do list. It's our statement of intent."

Cities around the United States are already moving to implement certain aspects of a Green New Deal, eliminating fossil fuels from their electricity systems, moving to electrify mass transit vehicles, and requiring new homes to have electric heat pumps. These are great signs of progress, but without the power and resources of the federal government we can't scale it up fast enough. Hence, it's too late for moderation.

I said, "Forty years ago, Shell and Exxon's own scientists warned them that what they were spewing into the atmosphere had the potential to cause irreversible damage. Had they not lied and hid that from the public, we could have had a moderate, incremental transition to renewable energy."

"Yes, and I am a moderate and I am here and that's the reason," Sam said. "Because the time to do this in a gradual and moderate way is past, so a moderate person wants to hurry up because that's the only available choice now."

Sam turned and looked directly into the camera. "So, whatever your temperament is, whatever your inclination is, this is the definition of moderation now, taking these different kinds of very bold steps. What looks like radical today is actually moderation, given the size of the problem itself and the time we have left in which to slow it down."

Sam added, "The insidious thing about this is that the catastrophe is being experienced differently in different places and that makes it possible for a privileged person like me to deny that it's happening. 'I'm okay. There's a little trouble over there maybe, but we're all right.' But what we need to do is exercise our empathy

Sam Waterston speaks.

and our imaginations. We have to be able to see ourselves in the typhoon victims in Japan and people that are being driven out of their homes because of climate change in Latin America. We have to see that their problem is our problem. This isn't just altruism. It's enlightened self-interest."

"Right," I said. "And

because of the heat already baked in, it's going to get worse before it begins to slow down. Hence, one of the things that the Green New Deal talks about is doing everything we can to prepare for the really challenging emergencies that are going to happen no matter what we do. And a country whose people feel safe and respected, people who are paid a decent wage, communities that know they are not considered sacrifice zones, are going to be much more able to withstand what's going to come at them in terms of extreme weather. The Green New Deal will grow resilience in people and in communities." Five months later, the coronavirus provided a tragic lesson on the importance of being prepared.

I was exhausted by the effort it took to make our first teach-in work in spite of the challenge of having two of us non-experts in the Green New Deal. I fled home to my hotel room and fell into a deep sleep.

The Rally

The next morning was the second Fire Drill Friday, and it was a very different scene from the prior week. The rally area was overflowing with press, paparazzi, and more people.

Onstage I opened with a brief history of the 1930s New Deal, because I wanted to remind people of this country's ability to rise to the occasion when faced with a monumental crisis.

WE'VE DONE IT BEFORE:
THE NEW DEAL

The monumental circumstances that created the original New Deal came about at a time of tremendous social unrest when the Great Depression threw a quarter of the population out of work. It was a time of labor protests, social protests, and riots about economic insecurity and inequality. People demanded that the government step in to alleviate these hardships.

The Depression was also a time of environmental collapse and mass migration. Fossil fuel companies encouraged midwestern farmers to shift to mechanized plowing, which disrupted native grasses that held the topsoil in place. When drought hit the region in the 1930s, high wind and choking dust killed livestock, and crops failed on the overplowed land, forcing 2.5 million people to flee. My father, Henry Fonda, starred in the movie **The Grapes of Wrath** about that crisis.

Angry, organized citizens, workers, and farmers made their demands known to Roosevelt, and what he said to them is important for us to keep in mind today. He said, "I agree with you. Now go out and make me do it." Pressure from citizens made him do it, and in order to lift the country out of its deep suffering, the president launched the New Deal.

In his first hundred days after he was sworn in as president, Roosevelt created government programs that put many millions of people to work in hundreds of public projects all across the country.

The Civilian Conservation Corps employed three million young men to restore the Great Plains; the Works Progress Administration hired millions to construct public buildings and roads like the dams and waterworks that brought power to the South through the Tennessee Valley Authority. This massive outlay of money for the public good built America's middle class.

It wasn't perfect. In order to get southern Democrats to support it, Roosevelt cut African Americans out of much of the New Deal, and women were largely ignored. We must learn from those mistakes and do better going forward. But the New Deal got so much right, and we benefit from that today.

The rich and powerful hated the New Deal because it set a precedent for the federal government to play a central role in the economic and social affairs of the nation. It was criticized as fascist or socialist; bankers tried to overthrow Roosevelt. Big Business, Big Railroads, Big Banks ranted and raved against it, but there were millions in the streets demanding that Roosevelt do even more because it was helping them, and because of that it succeeded.

This is exactly the kind of brave leadership we need to see from our next president, and there are very smart policy experts putting together a first-ten-day plan for that president, laying out things he can do through executive order without waiting for Congress. This will be a huge, disruptive, super-ambitious undertaking, and yes, it will cost a whole lot of money. But the cost of inaction is huge. Over the last three years, the total cost of billion-dollar weather and climate events exceeded $450 billion! And the COVID-19 pandemic has shown us that government can find the necessary money in an emergency.

The same interests that hated the New Deal are the ones today telling us that the Green New Deal is bad, that Big Government is bad, and they've convinced a lot of people of this, the very people who are actually hurt because government is being shrunk and taken over by corporate interests.

The fact is, the policies proposed by the Green New Deal are in line with what the American people have done before . . . because there was no choice.

There is no choice now.

"Make no mistake, change is coming, by disaster or by design," I said. "The Green New Deal provides the design to bring us all into a sustainable future."

Most of our speakers that day were young, and it wasn't until I heard them that I realized the extent to which many are traumatized by the prospect that the future they will inherit may well border on un-livable and are resentful that they must sacrifice their time, their educations, their chosen professions even, to fight for a future. I was reminded how the Green New Deal gives these young people, specifically, a vision of that future worth fighting for.

The first person to speak, Abigail Leedy, was an eighteen-year-old organizer with the Sunrise Move-ment. She told the crowd that a month earlier she found out she'd been accepted to college on a day when she was part of a sit-in at a congressman's of-fice to demand that he support the Green New Deal. When she thought about what was happening to the climate, how the world was on fire, she decided she had to forgo college to fight full-time for a Green New Deal.

She spoke from personal experience about how, in her hometown of Philadelphia, summers now are the hottest on record, and the school district can-celed classes for six days in her senior year because the temperatures were so high it feared it was not safe for the students to be inside the un-air-conditioned classrooms.

"In Philly over a hundred people have died in the

last ten years during heat waves. They send us home from school, but most kids go back to homes without air-conditioning and neighborhoods that are 10 degrees hotter than in the suburbs," she said.

Abigail described an oil refinery situated in a densely populated poor neighborhood where large numbers of children suffer from asthma. That refinery exploded in 2016, raining down debris on the homes of South Philly. Although no one was killed, the disaster released five thousand pounds of hydrofluoric acid into the air.

"In Philly people die because of fossil fuels, because we don't have nurses in public schools, because they are poor. Young people are born poor, stay poor, and maybe die later from a heat wave or a fossil fuel explosion that they had no role in creating.

"My friends have a running under-the-green-new-deal joke: Every time we think of something so good it feels impossible, we make the joke. Under a Green New Deal, corporate bookstores are public libraries. Under a Green New Deal, we're going to turn all these gas stations into parks. Under a Green New Deal, all the public schools have air-conditioning. It's mostly a joke, but it's also not."

I found Abigail's words heart wrenching.

Jasilyn Charger spoke next. She was one of the indigenous youths who had run two thousand miles from Standing Rock, North Dakota, to Washington, D.C., in 2016 to deliver a petition with more than 140,000 signatures to the Army Corps of Engineers asking

Jasilyn Charger speaks.

them to stop building the Dakota Access Pipeline through their lands and waters. She, too, questioned why young people have to be on the front lines of the climate crisis instead of "living our lives as young people. It's because we feel the fear in our hearts and souls that our future is threatened. But today, just to see that we're supported by the young and the old, by the rich and the poor, standing united as one, on a field in front of the Capitol building brings so much strength to my heart."

The next speaker, Charlie Jiang, a twenty-four-year-old climate campaigner at Greenpeace USA, had wanted to be a physicist to uncover the mysteries of the universe and had expected that he would lead a comfortable life.

"But in the face of a crisis where so many of my friends and family are suffering from pollution and fire, comfort is no longer an option for people like me because first and foremost, as global citizens, we

Charlie Jiang speaks.

have a duty to each other and I had to ask myself, how could I look at the stars as a physicist when so many people are suffering on this planet?"

Charlie said a new age of a Green New Deal was possible if millions of people come together, take to the streets, vote, and engage in civil disobedience.

"The next decade is going to be crucial, and the next year will be crucial. We're going to need all hands on deck in 2020 to take bold actions so that we can elect leaders who will be champions for a Green New Deal and pass policies to end the era of fossil fuels."

I concluded the rally imploring people to vote. "Make sure that you're registered to vote. Make sure that everyone you know is registered to vote. When the time comes to vote, reach out to people who may not be able to get to the polls and drive them to the polls. Every vote counts. And I'm not just talking about the president or the Senate or the House but state legislatures, governors, boards of supervisors, every one of the elected officials that you can vote for can make a difference. And before you vote for them, find out how they feel about the climate crisis. Make sure that they understand the urgency that's involved because who we elect next November is going to be so important. And then you have to mobilize. Think about the original New Deal in the 1930s. Realize that what got Roosevelt to do what he did, that really brought this country out of despair in desperation, was because people were in the streets. They forced him to do it. Demonstrate. Organize. Vote."

Our march from the rally to block the intersection between the Capitol and the Supreme Court was chaotic as reporters pressed in, shoving microphones in our faces. It seemed as if they were five-deep on all sides. Most were friendly, but there were also the "Jane, did you take a plane to get here? Don't you think that's hypocritical?" and "Hanoi Jane, why do you think you can speak about climate?" I refused to engage. My daughter, Vanessa Vadim, surprised me by taking the train down from her home in Vermont to be with me. One of my closest, longtime friends, the film producer and curator of the Tribeca Film Festival Paula Weinstein, had come, too. They were so buffeted about by the cameras pressing in from all sides, it was hard to keep moving ahead, in a line, chanting and holding the six-foot-long "Green New Deal" banner.

All of them, Vanessa, Paula, Sam, and the actor Katherine LaNasa along with twenty or so others, planned on engaging in the civil disobedience. I was supposed to receive the Stanley Kubrick Britannia Award for Excellence in Film that Friday, but I knew I couldn't attend the ceremony because Fire Drill Fridays were my priority. As I was being led away by the police with my hands in ziplock cuffs, I was filmed shouting, "Thank you. I'm honored!"

Apparently, the clip was enthusiastically received at the gala event. I hoped it would inspire more celebrities to join. Whether we like it or not, our society is very celebrity focused, and having a famous person publicly join a movement helps bring the press out and expand the reach of the message. The passion and life-changing commitment of the students moved and inspired me deeply. By bringing celebrities together with these youth leaders and climate activists, Fire Drill Fridays was helping to amplify the voices that needed to be heard.

This time, because there were more of us, instead of being held in the cells inside the police station, we were taken to a large, cold warehouse used for processing by the police for these types of occasions. I kept an eye on Sam, who was chatting with some of his fellow detainees and seemed in fine fettle. I knew Paula would be okay. As an anti–Vietnam War activist in the 1960s and 1970s, this was far from her first encounter with the Capitol Police.

I met interesting people during our three hours of detention: a manicurist from Delaware who had heard about Fire Drill Fridays on the radio and had come

A MODERATE'S VIEW
OF CIVIL DISOBEDIENCE

Sam Waterston

"I came out of getting arrested a different person than I went in. Pieces of myself came together that were being held apart by dread of—if I faced it—how much bigger than me the climate problem was going to be and denial, the old delusion that if you stick your head in the sand, the problem will go away or, at least, not land on you. As I've said before, I'm not a radical. In a radical situation, though—and the climate emergency is a radical situation—radical action is the **moderate** thing to do. To even feel a little bit whole, we need to do something out of the ordinary about it. Getting arrested over any cause was completely new to me. So far in my lifetime, there have certainly been other issues that merited it, but the climate emergency is at the top of any list. Of course, we all need to be modest about what we as individuals can accomplish, but getting arrested might actually work! Everyone should try it. As always, Abraham Lincoln hit the nail on the head: 'We must think anew, and act anew, and then we shall save our country.' That's what it feels like to get arrested over the climate emergency: thinking, and then acting, anew. I can't recommend it highly enough."

with her daughter, neither of whom had ever done anything like this before; a professor of rhetoric from a midwestern university who had a bit of Diane Keaton humor in her; and a Pakistani woman who works to protect Muslim refugees from ICE.

It was still daylight when we paid our $50, gave our thumbprints, and were released. There was a contingent of young climate strikers along with our jail support team waiting for us. I could see that Sam had been affected by the whole experience in a positive and deep way.

A few weeks later, Sam, the self-described moderate who had never engaged in civil disobedience or gotten arrested until our second Fire Drill Friday, had joined the students protesting their university's fossil fuel investments during halftime at the Harvard-Yale game and gotten arrested on the field. He sent me a photo someone took of him in handcuffs looking up with a big grin. "Now look at what you've done!" And Sam came back to D.C. with his sister Ellen for the twelfth Fire Drill Friday and did it again.

Vanessa was off to visit some friends, so Paula and I joined Jerome at his weekly climate strike at the While House.

I could feel my voice starting to weaken, so I left the chanting to Jerome and was amazed at his vocal stamina and persistence. He stood there, shouting climate chants for at least thirty minutes while I engaged with people who stopped by for selfies and talked about the climate crisis until it was time for us seniors to call it.

It was the end of another week, another Fire Drill Friday under our belts. I could feel it was growing.*

What Can I Do?

The surge of public interest in a Green New Deal is so exciting! Finally, we're talking about a response as ambitious as the science requires. The Green New Deal resolution, introduced by Senator Ed Markey and Representative Alexandria Ocasio-Cortez, is just the start. There is much work ahead to fill in the details, build support, and then make it real. There are many ways to get involved.

With gridlock in the Senate and White House stalling meaningful federal action, we will likely see more progress on Green New Deal policies at the state and city levels. Many states and cities are working on Green New Deals tailored to their specific needs. The GND in Seattle—known for its terrible traffic—includes free public transit. California has a statewide GND act that proposes strong action on climate while increasing affordable housing and addressing homelessness, which is a crisis in the state.

Many local governments are advancing the GND,

*Please see Appendix B: "Civil Disobedience," by Annie Leonard.

even if they don't use that name. About twenty states and a hundred cities have plans to achieve 100 percent renewable electricity. So check out if there is a GND initiative—whether named GND or not—where you live. If not, gather friends or contact a local climate group to write to your city and state representatives to urge them to get going. Organize community dialogues to identify what is necessary to start a GND where you live.

The GND plan your city or state develops will be better, and support higher, if diverse stakeholders, including labor and environmental justice groups, are engaged from day one. The New Consensus has useful materials for building unanimity around a Green New Deal; see its website for the different components of a Green New Deal and build one that suits your community. The Sunrise Movement has loads of opportunities for young people to get involved locally and nationally, including a guide for starting a local group to work for a GND. Sunrise's website lists the U.S. Senate and congressional co-sponsors of the GND resolution: Check if your representatives are there. If not, write and call them—and recruit friends, too—to rally support.

Many communities are building on the Green New Deal: Indigenous activists with the Red Nation have a Red Deal that incorporates Native leaders' wisdom, ocean advocates are crafting a Blue New Deal to protect our oceans alongside a GND, and feminists have developed a set of feminist principles for a Green New

Deal. Discuss incorporating these ideas—or others—into the GND campaign you join or start.

When you're putting together a Green New Deal package for your community, don't forget transitioning off fossil fuels! That means stopping the permitting of new fossil fuel operations, responsibly winding down existing operations, and ensuring those whose fossil-fuel-dependent jobs get eliminated find good work elsewhere. A GND that advances equity and security for all has the power to steer our country toward true prosperity as we address the climate emergency.

Of course, we're not going to get a GND passed if we don't elect climate leaders at all levels of government. Ask every candidate—for president, governor, mayor, city council, zoning board, planning commission—**every single one**—if they support a Green New Deal. Let them know that you support climate leaders on Election Day—and every day! Ensure that friends, neighbors, and co-workers are registered, and then let's all get out the climate vote!

Ted Danson gets arrested.

CHAPTER FOUR

Oceans and Climate Change

The week leading up to our third Fire Drill Friday was probably, physically, the hardest week of all for me. On Saturday, I flew to California, where Lily Tomlin and I had contracted to do five speaking engagements in the Los Angeles area. On Thursday, I hosted the teach-in remotely via webcam and planned to take a red-eye back to D.C. so that I could arrive in time for the Friday rally. Red eyes and I don't get along very well.

This Fire Drill Friday was especially emotional. You see, the first two focused on why I was doing them, what our demands and goals were, and what we saw as the solution. This week I was studying the effects of climate change on our planet, starting with oceans. My stomach was in knots.

I grew up in Brentwood, California. For my first decade of life, before the emergence of freeways and smog, I could see every day from our house in the hills a wide, shimmering expanse of ocean that inspired my little tomboy self. I spent summers at the Santa Monica beach learning to dive under waves and bodysurfing with my brother. I have snorkeled on the Great Barrier Reef while it was still fully vibrant and thriving. I'll never forget my first plunge off the wing of a seaplane far out at sea where there were no other people except my son and daughter. The wild variety of colors and shapes so startled me I almost swallowed a mouthful of water. I've been scuba diving with my children over coral reefs in the Caribbean and with sea turtles and seals in the Galápagos Islands. I've fished the high seas with every important man in my life, starting with my father. I spent ten years with Captain America, Ted Turner, who sailed every ocean on earth and won the America's Cup race. His respect and love for the ocean was contagious. The idea that my grandchildren won't have coral reefs to snorkel over because of climate change breaks my heart. Learning how much damage we have already done to the ocean, and how that destruction is accelerating, made this Fire Drill Friday urgent and personal.

Although I have loved the ocean all my life, I found there was a lot that I did not know about it, and I was ready to learn from our Fire Drill Friday experts.

The Teach-In

John Hocevar, the oceans campaign director for Greenpeace, joined us for our teach-in, and what he told us was eye-opening.

"The ocean absorbs 93 percent of the heat we generate and roughly 40 percent of the carbon dioxide we produce," he said. "In fact, oceans are our biggest allies in the fight against climate change. What will happen if we are not able to reverse the damage we've done to them? Our climate crisis will continue to accelerate.

"People have eaten roughly 90 percent of the world's big fish, which means that it's difficult for fishing companies and seafood businesses to turn a profit. So many of them have cut costs by not paying their workers a living wage. Some have stopped paying their workers at all."

John told us that they are seeing human trafficking, forced labor, debt bondage, and slavery on fishing boats.

"Besides the human labor abuses and the issue of overfishing," John went on, "there's the problem of the bycatch, which includes accidentally, or sometimes not so accidentally, catching a million sharks and hundreds of thousands of sea turtles that die in fishing gear each year. An important organization advocating for oceans, Oceana, estimates that overfishing, bycatch collection, and the demand for shark fins for soup kill 100 million sharks a year.

John Hocevar speaks. Behind him is Ted Danson.

"Did you know that for three billion people about 20 percent of their essential protein comes from the ocean? Did you know that about half of the oxygen we breathe comes from the ocean? Actually, that oxygen is generated by microscopic plants known as phytoplankton. These vital plants are one of the first casualties of the way climate change is acidifying oceans.

"As we burn coal, oil, and natural gas, carbon dioxide goes up into the atmosphere and warms our planet, causing climate change. That's global warming. But a lot of that carbon dioxide is absorbed directly into seawater, increasing the acidity of the water. That added acid makes it difficult for organisms like clams and oysters and coral reefs to form calcium skeletons.

"Possibly the most serious consequence of all," John said, "is its effect on plankton and phytoplankton, the small creatures that comprise the bottom of the food chain, feeding everything from tiny fish to giant whales. Most plankton, including the oxygen-producing phytoplankton, can be damaged by the acidification of the ocean water. They're already starting to be degraded, and some have disappeared.

"What's going to happen when we lose whole classes of plankton?" John asked. "We don't know. But our oceans are already producing less oxygen than they were just a few years ago. We're conducting an uncontrolled experiment in the open oceans at a global scale, and it's not okay.

"Offshore drilling is a big part of what's driving climate change right now," John continued. Offshore drilling degrades the environment, and oil spills and burning of oil extracted from the ocean are big factors in climate change. John described the direct effect one oil spill had on wildlife. "After the Deepwater Horizon explosion in the Gulf of Mexico, I was there on the Greenpeace ship **Arctic Sunrise,** working with scientists to determine the scope and impacts of the disaster. What I saw, the impacts on seabirds, upon dolphins, on really everything that swims and crawls in the Gulf of Mexico, gave me nightmares for weeks. And again, this is not okay. We can't just let this keep happening."

At the teach-in, we were also lucky to hear from Ted Danson, who has spent decades of his life fighting for the world's oceans. It's unusual for a major celebrity to be somewhat of an expert on something like oceans, but Ted Danson is, which is why I scheduled this particular Fire Drill Friday at a time when he was free to join.

"Basically, one-third of the world's catch is being thrown overboard," Ted said. He explained that seventeen years ago, in 2002, when he merged his American Oceans Campaign with Oceana, where he now serves on the board, they focused on the serious problem of overfishing and bycatch. "Fishing boats used to have to lift their nets over rocks, not to protect these fragile fish habitats, but because they didn't want to tear the nets. But when their engines got more powerful, at least for the large trawlers, they had enough power to drag their huge nets on rollers along the bottom of the sea, destroying the seafloor habitats, turning it all into gravel pits. And, as John said, when they pull these enormous nets up, they are filled with all kinds of sea animals that they were not fishing for, which they throw overboard.

"Our message at Oceana became 'Save the Oceans, Feed the World,'" Ted said. "We work with the fishing management systems in the twenty top coastal countries and get them to start fishing sustainably. The people who have investments in the huge trawlers fight us, but the locals understand that if you manage fisheries correctly, there'll be more fish, more money, more jobs, more food. That's been proven time and time again. And most of the fisheries and the nurseries are in that first two hundred miles of every country's economic zone."

John reminded us that though the coastal waters were important, to save the sea turtles, the whales, the seabirds, and the tuna, we had to take care of the high seas, too.

"These are all species that migrate across the world's oceans. And right now, unfortunately, the high seas are poorly regulated, basically unprotected, and it's the Wild West out there and there's pirate fishing. We're going to lose things in our coastal waters even if those are well managed."

"Come on, John," I pleaded, "give us some good news. I feel like putting my head under a pillow and crying."

And he did. He told us that ocean advocates have had some real success stories recently and that in the summer of 2019 more than 190 nations came together through the United Nations to start negotiating a Global Ocean Treaty to establish ocean sanctuaries that will help preserve and grow ocean life. Sanctuaries are the most proven, cost-effective tool to rebuild depleted populations of fish, to protect biodiversity, and to increase the resilience of species and ecosystems so they have a better chance of surviving the impacts from climate change, plastic pollution, and ocean acidification.

"If we get this right," he said, "and we have to get this right, this treaty is going to enable us to scale up ocean sanctuaries that actually will be able to fully protect 30 percent of the world's oceans from extractive use for the first time. Right now, we have about 2 percent of the oceans fully protected and about 5 percent partially protected, but scientists say we really need to get to 30 percent or more by 2030."

Ted shared some more good news about protecting fish stocks. Oceana has teamed up with Google and a satellite company called SkyTruth for an oceans surveillance program called Global Fishing Watch, partially funded by a generous donation from Leonardo DiCaprio. Every ship over three hundred gross tons traveling on international waters, ships that can be three hundred feet long and are capable of processing 350 tons of fish a day, has to have a device that lets a satellite know where it is and where it's been. When a ship sails swiftly through a protected marine sanctuary, you know it's going someplace else to fish. But if all of a sudden it slows down and zigzags, you know that the ship is probably fishing illegally. By notifying someone onshore when the ship docks, authorities can search it and hold the wrongdoers accountable. It's tricky, though. Some of these ships get refueled on the high seas, where they off-load their often-illegal catch onto another boat. If a search of the ship onshore determines that it is not just fishing but engaging in human trafficking, drugs, and gun smuggling, activists can take this information to the ship's insurance company and ask, "You know that ship that you insure? Here's proof of all the illegal things taking place on board. Are you sure you want to insure that?" Without insurance, the ship cannot legally operate.

I learned so much during the teach-in. I had not known the role the ocean plays in generating the very air we breathe. Learning what business as usual has done to this and other parts of the ocean made it hard to sleep. I would lie in bed having conversations in my head with the members of the Senate who deny or delay on climate, feeling impotent because I couldn't find the words that would make them understand what's at stake and persuade them to do something.

Could they ever understand what it means that the Antarctic peninsula is warming faster than almost anywhere else on earth except the Arctic, which is warming even faster? Did they know that as we lose that sea ice, we're losing the population of krill, tiny shrimplike crustaceans that are critical food, the energy source for penguins and whales and so much else?

Could I convince them of how important it is to protect the whales because of their vital role in fighting climate change? One whale can sequester thirty-three tons of carbon over its lifetime, which it does by pooping. As whales swim, they eat plankton, which absorbs carbon, and when they turn that into poop, the poop sinks to the bottom. When they die, their bodies sink to the ocean floor, taking that carbon out of the atmosphere and sequestering it at the bottom of

SHOULD WE BE EATING FISH?

Ted Danson

"I don't think we need to stop eating fish. But we need to be educated about what fish we shouldn't eat: the big billfish, swordfish, king mackerel, sharks. The top feeders. You shouldn't be eating those fish, because they have so much mercury in them from eating little fish further down the food chain." Another reason not to eat the larger fish is that they have a longer growth cycle before they reproduce. When you eat a bigger fish, you might have interrupted that species' reproductive cycle, depleting the fish population over time.

"The phrase that Oceana uses is eat small, local, and wild. Sardines, anchovies. There are a lot of wonderful fish you can eat that are not dangerous to your health. But farmed fish is tricky. You don't want to eat farmed tuna or salmon because they are carnivorous, and it takes about five pounds of ground-up wild fish to feed and make one pound of farmed salmon. Many of the small fish fed to farmed fish are ones that we could be eating directly, making farming fish inefficient."

the sea. So throughout their lives, and even when they die, they are taking carbon out of the lower atmosphere.

If only our government representatives understood that if we were able to rebuild populations of whales to the point where they were before we started killing them, we would be able to capture about 1.7 billion tons of carbon per year. In spite of the scientific consensus, many of them still deny that carbon emissions are a problem.

At the very end of the teach-in, Ted had said something that surprised me. He said that when he had turned seventy, he thought it was time to slow down, relax, and enjoy. "And then you called and told me what you were doing in D.C. to confront the climate crisis and you asked me to engage in civil disobedience with you and risk arrest. The idea of getting arrested somehow focused my brain a little bit, and I began to realize that everything we've done, that people all over the world are doing, to make the oceans vital and healthy, can be undone because of climate change, the acidification and warming of the oceans and the harm that plastics are doing, and that's related to oil as well . . . it all comes back to fossil fuels. I'm starting to think about all this in a new way."

This was only our third Fire Drill Friday, and already Ted Danson, big influencer, was thinking in a new way. Do you have any idea what that meant to me?

The Rally

At the Fire Drill Friday rally, Jennifer Jacquet, an associate professor of environmental studies at NYU specializing in marine ecology conservation and evolutionary biology, encouraged us to think of the animals around us, on land and on sea, not just as commodities. "Aquatic animals are the last wild animals hunted at a global scale for food. There are at least one trillion fish killed every year by fisheries. We don't think about fish as species; we think about them as seafood," she said. She noted how fish are moving toward the poles at a rate of twenty-six kilometers a decade to avoid the warmer ocean temperatures. This means large numbers of fish will be disappearing from the tropics by 2050. How fish were behaving as their environment changed was telling us something about the future.

Jennifer Jacquet speaks.

Whitney Crowder, Ted Danson, Jane, Maddy Carretero (behind Jane), John Hocevar, Jennifer Jacquet, Denise Patel, and Laura Flanders (behind Denise) pose for a shot in front of the Capitol before the rally starts.

"Some say the movement of fish represents opportunities for fisheries in the Arctic, and some worry about the future geopolitical conflicts over these fish. They speak about these animals merely as resources, instead of considering that this is nature sending us a signal," she said. "Corals are bleaching, and fish are moving because we have made their world uninhabitable. They are moving so they can breathe. Climate change is just one threat to aquatic animals. The largest threat is still direct exploitation, the global fishing industry. Factory fishing has to end."

The sea turtles face a different consequence of global warming, one that is painful to hear about.

What I learned that day about them I will carry in my heart forever. As I said at the start, I have snorkeled with sea turtles, getting right up close and looking them in the eye. Kind, sorrowful eyes, they have. The reptilian version of a golden retriever. John Hocevar had already told me that because the sex of sea turtles depends on the temperature of the egg, due to warming seas practically all sea turtles in Florida are being born female. Almost no male baby sea turtles born in Florida for years. Thus no reproduction.

I asked John Hocevar to be sure to say something at the rally to give us hope as well as what we can do. He did just that, saying, "I have to fight getting depressed about the damages being done to the oceans, but taking action, striving to make positive changes for the things I care about, helps with that and it can for you as well.

"It's time to make a scene. The science is clear that our oceans are in trouble, and that **we** are in trouble," John said, urging everyone to get more active.

"It's important for us all to do what we can as individuals. But our real power comes when we work together to ensure that politicians and corporations understand that we want better options. We want a world where the choices we have available to us are not going to destroy us. When you succeed, spread the news and inspire others to do the same, and then take it to the next level, and work on your town or city, your county, your state, and your country. I promise you, you won't be alone."

THE PLIGHT OF SEA TURTLES

Whitney Crowder

Whitney Crowder is a sea turtle rehabilitation coordinator at Gumbo Limbo Nature Center in Florida.

"Sea turtles are dying in record numbers, and it's directly because of us," Whitney said. She described tiny turtles starving to death, their stomachs too filled with plastic to ingest food. Other sea turtles died entangled in discarded fishing nets or weighed down by our trash. Pollution-related diseases caused tumors to cover their bodies, including their eyes and their flippers, making them blind and unable to swim. All of this she had seen firsthand.

"Every year due to strong winds, hundreds of these little sea turtles are washed back to shore along our Florida coastlines and many of them die," she said. "This year we have found 100 percent of our dead, washed-back sea turtles with stomachs full of plastic. Every turtle is filled with plastic. In one dead baby sea turtle, we found over 104 pieces of

Whitney Crowder speaks. Behind her are Ted Danson and Denise Patel.

plastic in its stomach. These hardy sea turtles, who survived the mass extinction of the dinosaurs, were being wiped out by our plastic trash, which we see clogging the surface of the water. Our plastic trash that flows into the ocean is broken down by the sun and the saltwater into smaller pieces that the sea creatures ingest. We used to think this was just a problem for the creatures that live in the waters close to the surface, but in fact it filters down all the way to the seafloor, creating hazards for sea life all along the way.

"Losing sea turtles, we create a cascade of destruction in our oceans that we may never recover from," Whitney said. "Sea turtles are an indicator of the health of our oceans. Without healthy oceans, life on earth cannot exist. Our oceans can't wait. I never felt like we're running out of time to make a change until now. The sea turtles are warning us now that they are in trouble and so are we. Is it possible that by saving the sea turtles, we can save ourselves as well? We have the power to turn this around, and I think we can because the future of our children will depend on the actions we take right now."

Ted Danson and I joined many others to engage in civil disobedience. I caught sight of his face as he was handcuffed. He was radiant.

It was still daylight when we were released into the care of jail support. I got out before Ted, and when I went to hug and thank him, he took my face in his hands and said, "No, thank you, Jane. This has been important." And the way he said it, and the way he made me look him in the eye, let me know that something had shifted in Mr. Danson.

And it wasn't just the celebrities who spoke to me about the effect the actions had on them. Many others for whom civil disobedience was new told me, "I won't be the same." I have thought a lot about this. I think people's reaction has something to do with what happens when you make the decision to put your body on the line for something you believe is right and important. You aren't sure what's going to happen, but you know you will be arrested and for a period of time you have no control and you go ahead and do it. For me, anyway, it's like stepping into wholeness. You align your body and your values, and that's something we don't get many opportunities to do these days.

What Can I Do?

Whether we live near or far from the world's oceans, we can help protect and restore ocean life. To start, simply eat less fish! Reduce your fish consumption or avoid eating fish not only for the oceans themselves but also to protect the people who rely upon fish for their sustenance and livelihoods, such as island communities. If you need to eat fish regularly, try to eat smaller—and no less delicious!—fish that are lower on the food chain, such as anchovies, sardines, herring, and mackerel. For a regularly updated resource on which seafood to eat, check out Seafood Watch (seafoodwatch.org).

Fish markets and grocery stores need to hear that their customers care about sustainability. Ask at the grocery store, seafood market, and local restaurants where they get their fish. Ask how the fish was caught, and if the overall population of that species is healthy. (You may want to review SeafoodWatch.org before you go shopping or share this resource with the folks you talk to.) Does the fishing boat use ground trawlers that destroy the ocean floor and kill tons of sea life as bycatch? Was it line caught by a local fisher? Farm raised or wild? Are the workers on this fishing boat, or with this seafood company, treated well and paid fairly? Can the retailers verify that the seafood they're selling was **not** caught by enslaved or human-trafficked people? If they don't know, encourage them to find out so they—and the customers—aren't complicit in wrongdoing.

The employee at the fish counter, or at the checkout stand, might not be able to answer these questions, or have any power to influence the supermarket chain's practices, so be respectful and polite. We know, however, that consistent inquiries from customers about the items a store offers can make their way to the top of the corporation and change things. If enough people ask, sustainability and ethical sourcing can become part of a grocery store's marketing, meaning something to brag about.

Because activities on land end up impacting the ocean, our personal choices such as how we commute, garden, and shop can help reduce pollutants making their way to the ocean, especially if lots and lots of people do them. As much as you can, choose biking instead of driving, stop using toxic pesticides and fertilizers in your garden, and eliminate as much plastic waste as possible, especially single-use plastic that is used once and tossed away in minutes.

If you're lucky enough to be able to spend time at the ocean, take pictures of the plastic waste you see on the beach or in the water. Capture the corporate logo of the company that made this waste, and let it know that you expect better. If you use social media, post the pictures and tag the companies in your posts so they know their customers are paying attention.

While changing our individual consumption habits can help, the greatest change comes from collective and political action. Ask city and school officials if they're using toxic pesticides or fertilizers, contributors to ocean dead zones where oxygen levels are too low to support life, on park and school grounds. Recruit neighbors and others to support phasing out poisonous chemicals. Organic landscapes are better for kids, better for groundskeepers, and better for oceans. Beyond your yard and neighborhood parks, ask your workplace to reduce or eliminate the use of toxic pesticides and fertilizers on company property.

Make sure your elected officials know you care about the oceans by writing to them to share information and urge action. This requires time, but thankfully you do not have to shoulder this burden alone. There are many organizations that support ocean life and advocate for policy change. Join Greenpeace, Surfrider, Oceana, or other groups that support ocean conservation, and ask your friends and family to do the same.

And above all, remember that healthy oceans help create a healthy climate! Everything we do to promote climate solutions in this book helps protect the oceans. That's great, because they need us, and we need them.

Rosanna Arquette and Catherine Keener join Jane onstage.

CHAPTER FIVE

Women and
Climate Change

I worked hard the first week of November, traveling to Detroit for a final speaking event with Lily Tomlin, doing interviews, researching for Thursday's teach-in, talking with my lawyer to determine what my next arrest would result in. Because I had a pending court date due to my three priors, I would most likely have to spend the night in jail. I was okay with that. I'd been in jail before . . . but never for civil disobedience.

I was sure that this week with its focus on women and climate change would be fun and lively. Many friends planned on joining, including Catherine Flowers, founder of the Center for Rural Enterprise and Environmental Justice, on whose board I sit, and my Big Sur pals, Rosanna Arquette and Catherine Keener, who had witnessed the development of my plan to move to D.C. for this climate action. I was excited for them to see how it was coming together. My daughter, Vanessa, was returning for a second time. My adopted daughter, Mary "Lulu" Williams, returned that day from two years with the Peace Corps teaching English in a rural village in northern Uganda and joined me at the Fire Drill. She would stay in D.C. with me for the remainder of my time. And my fierce friend, playwright and activist Eve Ensler, would be there.

For thousands of years, a patriarchal paradigm has ruled. It's the paradigm that has led to the climate crisis, an extractive, use-up-and-discard mentality that treats workers, those who are different, women, and the natural world as commodities, at men's disposal, for their enjoyment and their profit. Around the world, in countries such as Hungary, Brazil, India, the U.K., Turkey, the Philippines, Russia, even currently in the United States, we can see the apotheosis of this toxic mind-set in the nationalistic tyrants, strongmen, and would-be dictators.

Under the millennial-old patriarchal rule, the feminine principle has been not destroyed but suppressed. The spiritual teacher Eckhart Tolle, in his book **A New Earth**, says this has "enabled the ego to gain absolute supremacy in the collective human psyche." He says, however, that it is harder for the ego to take root in the female than in the male because women are "more in touch with the inner body and the intelligence of the organism where the intuitive faculties originate," have "greater openness and sensitivity toward other life-forms," and are "more attuned to the natural world." I like to believe this is true, but I know for sure that women have been socialized to be caregivers, more attentive to others. Perhaps this has something to do with why women tend to be less susceptible to the disease of individualism, are more conscious of our physical and spiritual links to the natural world, of our interdependence, of the importance of the well-being of the community at large, not just our small personal circle. Men fear that becoming "we" will erase the "I," the sense of self. For most women, our "I" has always been a little porous, whereas our "we" has been our superpower.

I think some of this goes back to our hunter-gatherer past. Men went out to try to spear animals and bring back meat. Anthropologists have written that on the occasion the hunter did bring back meat, which was often not the case, he would give it to his family or use it to curry favor with tribal leaders. It was the reliable food—tubers, nuts, and berries—gathered by women, young and old, that made up the family's daily nutrients. And if a woman's own family didn't need the food, she would distribute it to other tribal members. And if the younger women were pregnant or nursing, older women did the foraging. Grandmothers would also help with birthing, care for newborns, and were indispensable in advising the younger women about where the best water was, the juiciest berries, the poisonous insects. Survival meant respecting the interconnectedness between women. They truly depended on each other, and I believe that is baked into our DNA.

This is of utmost importance now because the climate crisis we face is a collective crisis that requires collective, not individual, solutions. And the challenge is that for the last forty years the idea of the collective, the public sphere, the commons, has been deliberately eroded and individualism has risen to take its place, but individually we are powerless to make needed systemic change. That's why individualism works to the advantage of the relatively few who wield power, and that's why we need to set aside our differences, unify around our common needs, because together is how we gain power.

According to Anthony Leiserowitz, a senior research scientist at Yale who studies public perceptions of climate change, the three countries where people are the least aware of the climate crisis are the United States, Canada, and the U.K. Why? Because, Leiserowitz says, those are the countries where individualism has taken root the most, especially in the last thirty years, fanned by right-wing news outlets like Fox and other Murdoch-owned media.

But even in those countries, as everywhere on the planet, it's women's sense of our interdependence that helps explain why we are the ones who save not just our own families but our communities during extreme weather events and what allows women to rise in greater numbers to face this collective climate emergency. As Gloria Steinem says, "Women are not better people than men. We just don't have our masculinity to prove."

These are some of the reasons that women are at the forefront of climate solutions. But in many ways, they also bear the brunt of climate change. In developing countries it's women who are responsible for producing 40 to 80 percent of food. They plant the crops, harvest them, fetch the water, and chop the wood, the things that allow their families to survive. But because of climate change, when crops are failing and water is scarce, women sometimes have to walk for days and still may not find these lifesaving resources. Climate change makes their job much harder.

And women make up 80 percent of climate refugees, people who are displaced because of extreme weather events, and they are the last to be rescued from those crises. Studies have shown that women are fourteen times more likely to die in a climate-related disaster than men.

What's more, women carry more body fat than men do. It is in that fat that a disproportionate "body burden" of fossil-fuel-based pollutants, pesticides, and chemicals is sequestered that can cause health issues such as cancers and can be spread to children in utero or through breast milk.

Here's something that you may not know that I learned that week as I studied women and climate change for our Fire Drill Friday. Reports show that there are significant increases in rape, sexual assault, and domestic violence in places experiencing climate-related disasters like floods and earthquakes or environmental destruction like mining, fracking, or drilling. When oil pipelines and fracking sites are under construction, it brings an influx of thousands of men into rural areas and on indigenous reservations where they are housed in "man camps."

In the tar sands of Alberta, Canada, and in the Bakken oil fields of North Dakota, there's been a surge in sexual violence against indigenous women. North Dakota has at least 125 cases of missing Native women, although the numbers are likely higher because records are not officially kept. Patina Park, executive director of the Minnesota Indian Women's Resource Center, said, "We can't be surprised that people who would rape our land are also raping our people."

The Teach-In

At the teach-in, Eve Ensler told us stories about what she'd seen as she traveled the world on behalf of stopping violence against women. She spoke of being in the Philippines right after Typhoon Haiyan, one of the most powerful typhoons ever recorded on earth. "Waiting around the devastated sites, men would literally pull girls out of the camps to disappear them into sex trafficking."

WHY OLDER WOMEN
ARE THE CORE
OF THE CLIMATE MOVEMENT

I've noticed every week that in the crowds that assemble on Capitol Hill for Fire Drill Fridays and among the folks who are arrested with us for civil disobedience, women outnumbered men three to one, and they tended to be older women. Studies I have done over the years have helped me understand why this is so. To summarize what is a multifaceted and fascinating subject, the question of why older women are best suited to take the lead in saving the planet: We have the time, the wisdom, the breadth of vision, and the numbers. There are more than forty million of us in the United States—51 percent of the aging boomers. Women are the single largest demographic in U.S. history. We have less to lose, and now we're not afraid to be angry. According to some anthropologists, with age we become the more assertive gender; it's partly a hormonal thing. And there's the future of the young—in some cases, our own grandchildren and step-grandchildren, nieces and nephews—to motivate us further. Does it seem overwhelming to you that older women could actually change how the world's institutions conduct themselves? Well, as I wrote in my book about aging, **Prime Time**, you need to think of a trim tab. A trim tab is a miniature rudder, a tiny thing that is attached to the edge of the big rudder at the back of an ocean liner. If you move the trim tab just a little, it can, with hardly any effort, build up a low pressure that pulls the rudder around. Women constitute a critical mass, and older women are the critical mass within the mass. Women over eighty-five are the fastest-growing age group in the world! Let's become, together, the trim tab on the rudder of the ship of state.

She told of being in Brownsville, Texas, at the border, listening to the stories of refugee women. There she spoke with a woman from Honduras who had farmed her husband's land, but a prolonged drought had devastated the crops. Her husband beat her for not providing the family with food. Gangs are getting worse because of poverty, and this woman was raped in the street. When she decided to flee, she was assaulted and raped again because there are no protections for refugees. She came to the United States for asylum and was put in a detention center where she was assaulted again.

"Because of Trump's return policy, she will surely get sent back over the border to Juárez, Mexico," Eve told us, "a place that is known to be a center of femicide. She'll arrive there without identity, without family, without anyone aware of her, without any place to go because Mexico's civil society is stretched to the max trying to support immigrants who are being sent back over the border. And guess what will likely happen to her? She'll get sent into the streets, into prostitution, and likely be disappeared and murdered. It began in Honduras with the drought, and you can see the step-by-step consequences of how climate crisis leads to sexual abuse, and I want to say our solutions have to be the same thing. We can't fight this battle separately anymore. We must stop seeing what happens to women's bodies as a different problem from what happens to the body of the earth. We have to now come together and we have to understand that when we fight for racial justice, when we fight for our bodies, when we fight for climate justice, when we fight for immigrant rights, they are all the same story that got separated at a certain point in time because patriarchy needs us to stay in silos, to not get that we are interconnected, interdependent."

Emira Woods, from Liberia, was also with me at the teach-in. She is a member of the international working group of the organization Africans Rising for Justice, Peace, and Dignity and was director of social impact at ThoughtWorks, a global technology firm.

Emira shared other stories of communities devasted

Eve Ensler speaks. On the stage behind her are Rosanna Arquette, Emira Woods, and Catherine Keener.

by fossil fuels. "Let me take you to a place in West Africa called the Niger delta. For decades, multinational oil companies like Mobil and Shell have been extracting oil in the Niger delta, leaving a wake of destruction in their path. There are serious health impacts from gas flaring, water contamination, and oil spills in community after community in the Niger delta. Along the way warriors have fought back, including the Ogoni people and the Ogoni leader Ken Saro-Wiwa, who was murdered in retaliation for his campaigning against the environmental and human rights abuses linked to fossil fuel extraction."

Emira wanted to make clear that women are on the front lines of the solutions. "Africa may be the epicenter of this climate catastrophe, but it's also the epicenter of the solutions, and it is women that are driving those solutions. Lamu Island is a beautiful area in Kenya right on the coast, one of the most biologically diverse areas in the world, and it's being destroyed because a coal plant is proposed to be built there. But the community is rising up, led by dynamic young women, saying, 'No, we don't think so.' They're courageously standing up to oil and gas and money.

"It is women in India that are green entrepreneurs, that are developing clean cookstoves to say, 'No, we don't need to cut down trees; we can have clean energy in a way that uses renewable energy, in a way that uses the power of the sun, that is readily available and affordable.' And it is women that are designing these new tools and techniques that are going to not only create an opportunity to bring light into the communities but also save the planet. And in environments where women are literally delivering babies by candlelight, these solutions are vital to saving generations to come."

Emira spoke about how women in Kenya, Tanzania, and Rwanda organized for nationwide bans on plastic bags; how in countries with higher numbers of women in parliament, they are more likely to pass environmental laws and ratify international environmental treaties.

Eve added, "I've always thought girls were the most powerful force. Look at Greta; look at all these young indigenous women, the young girls who ran from Standing Rock across the country to deliver a petition to the Army Corps of Engineers here in Washington. There are so many powerful girls right now, and they're going to lead us into the promised land in creative ways. They're using social media and culture, music, dance, art. They're using all these phenomenal tools to organize, to ignite, but also to touch people's hearts." She went on to tell us about a writer and healer, Deena Metzger, who has been talking about what she calls "extinction illness." She told Eve about the rise in attention deficit disorder, autism, suicide rates or suicide ideation, and depression in young people. "And there's a new syndrome that's developing which Metzger calls 'resignation syndrome.'

A STORY OF
WHY EVERY EFFORT MATTERS

Emira Woods

"There is the great Kenyan warrior, a Nobel Prize winner, Wangari Maathai. I had heard her tell the story of the hummingbird trapped in a forest that was being consumed by fire. While the other animals, more powerful ones like the elephant, stood still in terror, the hummingbird flew to the lake and took a little bit of water in its beak to drop on the fire and then raced back for another drop

Emira Woods speaks.

as the other animals looked on. The hummingbird said to them that she was doing the best she could. 'We must all be like hummingbirds,' Maathai had said. 'Using our power, using our voice to stop the pillage of the planet and to actually be a part of the solutions.' Like that hummingbird, Maathai planted millions of trees to stop soil erosion, provide shade, and absorb carbon."

"I think what's happening," Eve said, "is young people have no absolute sense of a future. And I think the solution to that is for parents to bring their children up to become fighters for the future. That is one of the ways we can overcome some of these disorders."

I thought of what young people like Abigail Leedy, Jasilyn Charger, and Charlie Jiang had said at previous Fire Drill Fridays, about the despair they felt that made them join the climate movement.

The Rally

It was a wonderful rally at the Capitol the next day, the first really cold one so far, yet people turned out in larger numbers than ever. It really hit me that day at our pre-rally briefing that Fire Drill Fridays had turned a corner. It was clear now that we'd touched a nerve, tapped into something that we had hoped was there but hadn't been sure of: Large numbers of people were feeling the need, many for the first time, to put themselves on the line against the climate crisis. Here's one example: a young woman named Alexandra Charitan, editor of an online magazine, **Roadtrippers**, who had learned about us on social media and come down from New York. When she came down again weeks later to risk arrest for the second time, she told me she had never done anything like that in her life and that it had changed her forever. "I'd always been the 'good girl,'" she said, tears welling up in her eyes, "always followed the rules. I needed to step away from that and assert who I am. And this Fire Drill Friday action spoke to me." I was elated, my heart swelling with the realization that if we could build out Fire Drill Fridays around the country, we could mobilize more and more people who would become activists, be inspired to vote for a climate president, and demand that she or he be brave like Franklin Roosevelt.

A couple of hours into our group detention in the cold warehouse, the police came and got me, thoroughly patting me down again, putting me into a police van where my plastic handcuffs were replaced by the traditional metal ones. I was shackled to the floor and driven to the D.C. Central Cell Block, where people are held overnight until their arraignment the next afternoon before a judge who would decide whether to dismiss the case or sentence the person to jail time.

When I entered the Central Cell Block around 6:30 p.m., there wasn't much going on. My feet were again shackled, and I was questioned about whether I had ever been sexually abused. I told them I had. "Were you ever abused while incarcerated?" the officer asked, to which I answered in the negative. There were posters everywhere, it seemed, asking if you had ever been sexually abused while incarcerated and, if so, to call a hotline. I didn't remember seeing such things the last time I was inside a jail in the 1970s, and I found it spooky.

Vanessa Vadim (Jane's daughter) gets
arrested. In front of her is Emira Woods.

Mary "Lulu" Williams (Jane's "adopted"
daughter) gets arrested.

I was photographed for my mug shot, offered a bologna and cheese sandwich on white bread wrapped in cellophane and a plastic cup of purple juice of some kind, and locked into a cell by myself with a female guard stationed outside all night to protect me. From whom? I wondered. The only people who could get in were other guards.

It was a narrow cell with two metal-slab "bunk beds" and a metal toilet. No sharp edges, nothing that could be broken off and used as a weapon or for self-harm.

It wasn't long before other prisoners began to arrive, but I couldn't see them. Hearing them was a different thing altogether. There was a man who banged on his cell door and walls all night long, emitting heart-wrenching, incomprehensible howls of anger and despair. Guards walked noisily up and down the aisle all night. I was about to ask them to please keep the noise down because I wanted to sleep but then realized what a sign of white privilege that would have been and I shut up, chuckling to myself as I imagined that was something Grace might do if this were a scene from **Grace and Frankie**.

I was more than aware that I was being treated differently because I was white and famous. A cell to myself, juice when I wanted it, a guard. I had heard just the week before how when Greenpeace's brave protesters had been arrested hanging from a bridge, blocking an oil-exporting channel at a site in Houston, Texas, the men and women of color had been badly mistreated. My friend Patrisse Cullors, co-founder of Black Lives Matter, has told me horrendous stories of how her brother, who suffers from schizophrenia, has been brutalized in prison in Los Angeles.

I did wall squats for a while, meditated, and eventually was able to sleep using a sweater as a pillow and my red coat both to soften the metal under me and as a cover. It almost worked.

Sometime the next morning, I was told to get my stuff together and was shackled and handcuffed again and taken by van to another building, where I was held for many hours in a clean white holding pen with a bench and, behind a low wall, another metal toilet. After an hour or so, four women were brought in. Two were eighteen years old, and the other two maybe in their late thirties. All were African American. One of the young women had her feet loosely bandaged and was moaning in pain. Apparently, she had spilled hot grease on them from a pan of chicken she'd been roasting. She said she had left her house in a car with her young child, no shoes, and driven away, scraping another woman's car in the process. She and the woman had gotten into an altercation, and she said the police had arrested her but left the child alone in the car, so she was understandably upset.

The oldest one, who was lying on the floor talking to herself, told me she'd been making breakfast and getting ready to go to work at Dollar Tree when she and her boyfriend got into a fight that led to her arrest for domestic violence. It was clear that all of them had been arrested before but that sharing a holding pen with an old white woman in a fancy red coat was a first. I told them I had been arrested for committing civil disobedience because I and a lot of others were trying to raise awareness about the climate emergency.

"Wait a minute, you chose to be in here?" one of them asked incredulously. "What's this climate thing about?" another one wanted to know. I did my best to explain what was happening to the planet, but it was clearly not of interest. I told them I was a film and TV actor and that helped me get the word out through the press.

They sat up and took a little more notice . . . but not all that much. Clearly all of them had more pressing things weighing them down. "What movies have you made?" one asked. I knew it was likely that they had seen one in particular. "I was in **Monster-in-Law** with Jennifer Lopez." Yep, that one all four of them had seen.

After a few hours, I was taken to another, larger holding pen with five other women. After a while, it became clear that none of them belonged in jail. They needed mental health care, decent jobs, freedom from abuse, a supportive community, and basic respect and fairness. One of the women was shivering, and I lent her my coat. The moment she wrapped it around her, she stood up straighter and held her head high, and I could see how beautiful she was. Had her life been easier, she could have been a model or, I am sure, had any number of other careers. The guards must have been watching and gotten concerned about how much we were all talking because, eventually, I was taken back to the smaller pen, where I waited alone. I felt badly that I asked for my coat back, but it had already become a symbol of our climate project and I felt I needed to hang on to it.

There was an elderly black woman now in the cell next to me. The whole time I was there, she was on an anguished, incoherent rant. I tried but was unable to make out what she was saying, but she never stopped talking. I felt she was having a psychotic break and wished I could have gotten her the help she needed. Too many people suffer from the effects of racism and poverty. Will we ever be able to get our country to do better at being more fair to start with, and providing social services and proper mental health care for those in need? It is something worth fighting for. I couldn't help thinking how all these problems will only be exacerbated as the climate emergency worsens; jobs, health care, social services will all be impacted.

By 1:00 p.m., I had been unshackled and released to the court. As I walked down the hallway, I could see the jail support team all there along with Emira Woods in her red Liberian robe and head wrap, the wonderful nuns Carol Gilbert and Ardeth Platte, who had been coming to all the Fire Drills and risking arrest, and my adopted daughter, Lulu. As I hugged

Brooklyn Decker (far left), June Diane Raphael, Abigail Disney, and Catherine Flowers stand at the banner with Jane as the group prepares to enter the Hart Senate Office Building.

them and scanned the faces, I caught sight of my daughter, Vanessa, and burst into tears, so moved that she had stayed over to welcome me out. She was crying, too. It was an important moment for us.

My lawyer told me that the D.C. prosecutor's office had "no papered" me, meaning that they had dropped charges and I could go. Again, another sign of privilege. My friends took me back to the hotel, where my daughter put me into a hot bath and then dressed my wounds. I still had deep, open wounds from surgery a month earlier to remove skin cancers due to my youthful years as a foolish sun worshipper, and they were starting to get infected.

That night I slept for thirteen hours.

What Can I Do?

We are never going to solve climate—or a whole host of related challenges—without women in leadership positions able to decide issues that affect their lives. The more we have women leading the climate movement, the stronger our movement will be. Countries where women lead embrace international climate treaties more often than those led by men. Paul Hawken's "Drawdown" study, which examined the top ways to reduce carbon in the atmosphere, found that educating girls and empowering women was one of the most effective climate solutions!

So, for those who identify as a woman, start by finding a community that can support (and hopefully join) you as you develop your climate activism, especially if you find yourself pigeonholed into more traditional roles. Those who do not identify as a woman, take steps to support the women-identified people in your lives. Take on your fair share of the housework and child rearing at home and logistical and administrative work in the office, freeing women to lead. If you are part of a climate campaign or organization, make sure you're not unconsciously limiting the participation of women. If there's not a good gender balance, find out what you can do to make this work more welcoming for women. If you want to advance equal pay and equal rights, the American Association of University Women is a good place to start.

Whether women have a **choice** about having and

raising children is critical for climate justice. Educating and empowering women lead to fewer unwanted pregnancies and more opportunities for women. The climate crisis disproportionately harms poor women and women of color, who are also most burdened with child rearing and other forms of care work. Learn more about reproductive justice and gender justice by joining organizations like SisterSong, Forward Together, National Domestic Workers Alliance, and One Billion Rising.

Unlike the United States, the vast majority of the world's farmers are women. Women farmers have proven to be better environmental stewards of the land. We have seen this in the Chipko, or "tree hugger," movement started by Indian women farmers in the 1970s and the Green Belt Movement that planted thousands of trees in Kenya, founded by the environmental activist and mother of three Wangari Maathai, who won the Nobel Peace Prize for this work in 2004.

Organizations like Food First and La Via Campesina work to protect women's rights to their land, including economic and civil rights, and safeguard them from sexual assault and violence. Educate yourself about the ongoing crisis of murdered and missing indigenous women, a disproportionate number of whom disappear near fossil-fuel-extraction and fracking sites across the North American continent. Raise awareness of this tragedy by supporting organizations like the Minnesota Indian Women's Resource Center. Encourage your friends and family to do the same.

Support the emerging young women-led organizations like Greta Thunberg's School Strike for Climate, Future Coalition, and Fridays for Future. Funnel resources into grassroots, women-led groups that focus on climate and gender equity and offer to get involved. Some of my favorites are Women's Earth Alliance, Women's Earth and Climate Action Network (WECAN) International, African Women Unite Against Destructive Resource Extraction (WoMin), and Women's Environment & Development Organization (WEDO).

Winona LaDuke's Hemp & Heritage Farm is another indigenous-women-led organization to support; this longtime environmental activist, who joined Fire Drill Fridays in D.C., is both resisting fossil fuel pipelines and growing hemp for renewable energy. Check out her innovative work, and mail her a donation while you're at it!

Elect more women to public office and other leadership positions, and make sure women are at the table when climate crisis solutions and environmental justice are being discussed, such as the Green New Deal. (There's even a Feminist Agenda for a Green New Deal, developed by women leaders around the world: FeministGreenNewDeal.com.) Be sure the women you're electing are committed to climate, social, and environmental justice. The British prime minister Margaret Thatcher's legacy is one of the worst for the environment and women's, workers', and immigrants' rights. Do your homework, and make sure women leaders know you're counting on them to do right by their gender, as well as the whole planet!

You can also encourage women to **vote**. The Yale Program on Climate Change Communication has documented that women are more concerned about climate and more supportive of government action than men. **So let's get women to the polls and vote for climate leaders!** And let's work to make sure there are climate-committed women running for election up and down every ticket around the world. Groups like EMILY's List have helped make this happen when it comes to reproductive choice. Organizations like the Women's March are focusing on climate and reproductive justice in 2020. Join in solidarity with sisters around the globe!

From left to right: Krystal Two Bulls, Jane, Ben Cohen, Phyllis Bennis, and Jodie Evans. They are marching to the White House.

CHAPTER SIX

War, the Military, and Climate Change

The second week of November, Ira and I took the train up to New York to do interviews about Fire Drill Fridays and the climate crisis on **The View**, Vice, and CNN. I had taken the train between New York and D.C. a few times in the past to get to demonstrations against the Iraq War and climate rallies, but it had become a regular thing since the start of Fire Drill Fridays. I loved having the three hours for quiet studying and writing. As we passed through working-class cities like Baltimore, Wilmington, and Newark, the degradation I saw out the train window was evident. I would envision how a Green New Deal would bring federal funds into cities like these to retrofit buildings for energy efficiency and to create green spaces.

This was also the week that a new and alarming statement about the climate crisis was put out.

"Scientists have a moral obligation to clearly warn humanity of any catastrophic threat and to 'tell it like it is.' On the basis of this obligation, . . . we declare, with more than 11,000 scientist signatories from around the world, clearly and unequivocally that planet Earth is facing a climate emergency."

Don Lemon surprised me by reading part of it on air during our CNN interview.

This was good. More people would be obliged to face the fact that climate scientists were unanimous in their view of the threat.

I was excited about the speakers who had agreed to come to Fire Drill Friday to discuss war, the military, and the climate crisis. I have to admit the links between war and climate weren't something I had spent much time thinking about. I was especially glad that Phyllis Bennis was joining us. I had met Phyllis in the 1970s when she worked with Tom Hayden and me to end the war in Vietnam. We had reconnected at a three-day gathering of movement leaders following Trump's election. Phyllis is a fellow of the Institute for Policy Studies in Washington, D.C., where her specialty is war, the military, and the Middle East. Being the ideal combination of academic and activist, she was my perfect tutor for this Fire Drill.

The Teach-In

"War and militarism create climate catastrophe, and climate catastrophe creates war," Phyllis said at the teach-in. One of the big reasons Syria erupted into conflict was climate: a three-year drought that drove 800,000 farmers from their lands into the cities, where they faced unemployment, economic inequality, and sectarianism. "It was all rooted in the three years of drought that were the effect of climate change," Phyllis said.

I was stunned when Phyllis told me that the military is exempt from most environmental regulations, meaning it can dump toxic chemicals and burn discarded gear without addressing the consequences to the environment. This may explain why 10 percent of the Superfund cleanup sites, sites designated as dangerously toxic and needing cleanup by the federal government, are on or near military bases. In 1991 and again in 2003 and beyond, our military fired depleted uranium in civilian areas in Iraq, leaving a legacy of elevated cancer rates for the people there and for our GIs for generations to come. We can see the link between oil and wars, the environment and wars, wars and racism, wars and Islamophobia, and all these things are rooted together in terms of the economic impact, the domestic impact, and the international impact.

During the teach-in, I asked Phyllis to talk about what people can do. She made a convincing argument

**Phyllis Bennis speaks. Ben Cohen and
Jerry Greenfield sit behind her.**

about why we have to cut our bloated military budget, which encourages us to take military solutions over less deadly ones.

Aha, I thought. I've had experience creating a movement to cut federal budgets. In the early 1970s, Tom Hayden and I and an army of organizers that included Ira and Karen traveled the country mobilizing people to call, write, and visit their Congress members to demand they cut the aid to the Thieu regime. That U.S.-imposed South Vietnamese government was so unpopular in Vietnam it couldn't have lasted were it not for American taxpayers' money propping it up.

After two years of this work, Congress cut the funds, and the war ended soon thereafter.

"Our military budget is enormous," Phyllis said. "Right now, in 2019, it's $719 billion a year. It's bigger than the next seven countries combined, which means Russia, China, Saudi Arabia, and the four next biggest armed countries.

"Fifty-three cents of every discretionary federal dollar of our tax money goes directly to the military. If you add in things like Homeland Security and the other militarization issues, it's more like sixty-two cents. It's a huge proportion of our money, and it doesn't keep us safe. We could cut almost $600 billion from the military budget, and what we would have left would be the equivalent of what Russia, Iran, Saudi Arabia, and North Korea spend **combined** on the military."

Ben Cohen and Jerry Greenfield, co-founders of the ice cream company Ben and Jerry's, were our celebrities that week because they are longtime activists against war and militarism. At the teach-in, Ben made a point that I feel is apt: Money spent improving the living conditions here at home is national security spending. That, not war, is what makes us safe. How secure is a country where children are being poisoned by toxic fumes in their homes, where we are ill-prepared for a pandemic, where too many can't afford health care and a decent education, and millions are homeless?

Phyllis provided facts that illustrated our out-of-control U.S. military spending. For example, the United States has eight hundred military bases all over the world.

"No other country has bases all over the world," Phyllis said. "I think Russia only has seven. Do we really need eleven aircraft carriers when other nations only have one?" Phyllis asked. "All this money spent on defense, does it make any of us safer?"

Phyllis quoted Martin Luther King Jr. to answer that. "He once said, 'A nation that continues, year after year, to spend more money on military defense than on programs of social uplift is approaching spiritual death.' And today I would add, 'approaching the physical death of the planet.' That's the urgency of now."

Krystal Two Bulls, an Oglala Lakota and Northern Cheyenne woman who is an environmental activist and a veteran of the war in Kuwait, was at the teach-in. She's also a member of About Face: Veterans Against the War.

"Krystal has experience on not one but two battlefields," I said in introducing her. "She served in Kuwait, but she's also fought against the encroachment of pipelines and fracking on indigenous reservations." Knowing our military's shameful history of slaughtering Native Americans, I was interested in Krystal's thinking when she enlisted.

She explained that she grew up believing it was important to serve her country and she was also steeped

Krystal Two Bulls speaks.

in the legacy of her ancestors, who were Native American warriors: Oglala, the Sioux, the Arapaho, the tribes that defeated Custer at Little Big Horn. She enlisted thinking that this was a perfect melding of her values, but once deployed, she began thinking more about the U.S. military massacres of Native people in Sand Creek, Wounded Knee, and Fort Robinson. It made her question what she was doing as a Native person serving her country in this way, although she kept those questions to herself because it was something that she couldn't reconcile.

When she got out of the army, Krystal dove into the environmental justice movement and organizing without ever mentioning to those she was working with that she was a veteran. This is when she made an important shift in what serving her country meant, from "being a soldier who defends the empire and is a tool of an empire and does what you're ordered to do without question, and the healing shift over to being a warrior who will do what is necessary for the benefit of their people and future generations and our relationship to the land. And now I see that my antiwar work and my environmental justice work are not separate things."

Krystal's unique perspective on the military produced some creative thinking about repurposing the skills soldiers learn when they are no longer enlisted. She hadn't planned on talking about it, but I asked her to. I know too many veterans who return stateside and feel lost, angry, and at loose ends because of the trauma of what they experienced and because the military does a poor job of helping them reintegrate into civilian life. I felt her ideas might inspire them. I know from my own experience, and from what veterans have told me, that activism, that healing shift Krystal talked about, can lift depression and give new meaning to life. I had certainly been inspired when ten thousand vets came to Standing Rock during the siege to support the water protectors.

Krystal said she thought the easiest and most obvious skills to translate to civilian life were those of military medics. Unfortunately, given how police in this country have been militarized, the climate movement, among others, is going to need medics. Protesters are being beaten and injured by rubber bullets and powerful water hoses. She also said that the military is expert in logistics, organizing, and coordinating large groups of people and getting them to target a specific objective.

"Handling logistics in the military, knowing how to manage these things, looking at the long-term goals versus the individual tactics that are being used, these are things people learn in the military," Krystal said. "Being able to bring movements together versus working in silos is important. I'm able to bring people together in those ways to accomplish whatever our goal might be for a particular campaign. Across the board for every veteran, every person in the military, that skill can be translated over to direct action, to civil disobedience, to organizing. Imagine what we could do with those skills."

As Krystal detailed her ideas, it was clear to me how important it is that women are at the table when solutions to the climate crisis are being discussed and designed, just as the voices from impacted communities should be at the table. They see things differently, draw attention to problems and solutions that others ignore, know what the real, nuts-and-bolts, on-the-ground needs and solutions are.

Phyllis objected to the frequent retort that if you're for cutting the military budget, you "don't support the troops." The truth is, she said, very little of this military budget goes to support the troops. In 2017, when she was working on the Poor People's Campaign, Phyllis found that twenty-three thousand active-duty soldiers and sailors were making so little money that they and their families qualified for food stamps.

"If you look at where half the military budget goes, it goes to military corporations and the CEOs of those corporations, who are among the best paid of all corporate executives," she said. "In 2016, the average salary of the top five military corporations, the CEOs were averaging almost $20 million a year, while active-duty service people were qualifying for food stamps." Shocking!

I welcomed the levity that ensued when one questioner from the audience asked Ben, "If Fire Drill Fridays were an ice cream flavor, what flavor would it be?" He answered, "I've been experimenting with hot cinnamon flavors. You know when we were kids, people used to eat Fire Balls and those little hot cinnamon

Ben Cohen and Jerry Greenfield speak together.

hearts. I think that's what the flavor would be. Fire Ball Friday."

After hearing everyone speak, I understood why the climate movement and the peace movement must work in tandem.

The Rally

It was the coldest day yet at the rally, the first time I needed my wool gloves. Jerry Greenfield, Ben Cohen's more reticent business partner, spoke. "We can do it. It is still a democracy. The people do still rule, and there is a presidential primary and a presidential election coming up. You know, we are people that care about each other, our brothers and sisters, not only in this country, but all around the world. We have no interest in bombing and killing and drone striking other people like us that are just trying to live. It never ceases to amaze me that the people in that building," he pointed to the Capitol, "follow what is being told to them by the lobbyists and by the money that goes into their pockets from the polluters and from the military-industrial complex that's causing them to represent those interests at the expense of all of us. We can make it happen. We can make our votes count, but it's more than just our votes. We need to be coming together in groups like this all around the country, and we need to keep on doing it just like Jane is doing it every Friday, a commitment, and I need to say, it's a multigenerational commitment. Let's go do it!"

At the end of our rally we were joined by the Poor People's Campaign, the antiwar veterans' group About Face, Veterans for Peace, CODEPINK, the Institute for Policy Studies, and an organization calling to impeach Trump for pulling us out of the Paris Agreement.

There must have been a thousand of us as we marched two miles down the middle of Washington's wide streets to the White House, holding our banners and posters aloft and chanting nonstop. The sheer volume of our voices made it thrilling. On the way, we came upon a smaller march of young DACA

The march to the White House. Krystal Two Bulls and Phyllis Bennis are on each side of Jane.

supporters on the sidewalk going in the opposite direction toward the Supreme Court, where a ruling on their plight was pending. The two groups stopped and acknowledged each other with shouts of mutual support. It was a perfect, spontaneous demonstration of movement interconnectedness and intergenerational solidarity.

We intended to engage in civil disobedience at the White House, placing ourselves on the sidewalk and loudly chanting, but the police chose not to intervene. Then we moved to block the western gate of the White House, where scores of us sat down and kept chanting, but still the police didn't arrest anyone, and because it was getting colder, we decided to disburse and get some Ben and Jerry's ice cream, which they had generously delivered to the protest site. Not bad . . . cold ice cream on a cold day.

What Can I Do?

Questioning U.S. military spending can be tricky, because people often see that as an attack on our soldiers, and on our country itself. Not so! The best way to honor our soldiers and our country is to stay out of unnecessary wars and use some of the money we spend on the military to make our country safer, more resilient, and more prosperous, which includes fighting climate change.

Because the military can be an emotionally charged

issue for many, be prepared when starting conversations. Here are some talking points to help you through the reactions you may get when raising the issue of reducing military spending.

Reducing the military budget is dangerous; if we don't have a strong military, the United States will not be safe.

As Phyllis Bennis taught us, the United States "could cut almost $600 billion from the military budget, and what we would have left would be the equivalent of what Russia, Iran, Saudi Arabia, and North Korea spend **combined**" on defense. Cutting two or three hundred billion dollars would reduce our military might somewhat but would certainly not make us less safe, and it would improve our ability to combat the climate crisis tremendously.

Second, the military is not doing such a great job keeping its soldiers safe. If you are a soldier or veteran, the U.S. government does not do much to help you transition to civilian life or provide the long-term resources you need if you have suffered physical and psychological trauma. U.S. military bases also leave literal tons of toxic waste, which the U.S. government does not require them to clean up, poisoning innocent civilians abroad and military personnel at home. And as the 2012 documentary film **The Invisible War** revealed, one in three women in the U.S. military will be sexually assaulted by their fellow male soldiers. Arguably, the U.S. military as it is currently run makes us **less** safe, but more on that below.

Cutting military budgets will cut military jobs.

For every military job eliminated, the federal government could replace it with better jobs in education, public infrastructure, health services, and more. These are jobs we desperately need to fill, with workers performing services our residents need. And besides, there is a hugely unfair disparity between the pay of CEOs for military corporations (averaging $20 million per year!) and that of the twenty-three thousand active-duty soldiers in lower enlisted ranks who earn so little that they and their families qualify for food stamps.

But what about national security?

As Ben Cohen of Ben & Jerry's ice cream reminded us, money spent improving the living conditions here at home is national security spending. That, not war, is what makes us safe. The United States has eleven aircraft carriers. Almost no other country has more than one. Are those extra ten aircraft carriers, which cost up to $13 billion **each**, really keeping us safe? Wouldn't we rather our fifty-four cents to every $1 spent on the military and militarism go to cleaning up the polluted air, soil, and water right here at home?

Or to improving our schools and increasing access to mental and physical health care? To vocational training and job placement programs for our workers, especially those who will be displaced by a transition off fossil fuels? Social services, education, environmental sustainability—these are the things that keep us safe, not aircraft carriers.

Okay, but what does militarism have to do with climate?

The climate crisis has **everything** to do with the military. Every single war fought in the past eighty years, if not longer, has been about oil. Going forward, global conflicts will be not just about oil for profit but also fights over increasingly scarce resources like water and food. To make matters worse, wars and related climate crises produce "climate refugees," civilians who have no choice but to flee their unlivable homelands yet can find nowhere that will take them.

The U.S. Pentagon alone is the biggest institutional user of fossil fuel in the entire world, which means the United States is responsible for the biggest release of greenhouse gases. The climate movement and the peace movement are one movement.

Learn more about how our cities, countries, or states support the military. You can find out about this at the National Priorities Project (www.national priorities.org). For example, you can learn how much you and your neighbors are paying every year for wars in Iraq and in Afghanistan. There's also a tool on the website (www.nationalpriorities.org/interactive-data /trade-offs/) that shows you how many kids you could educate at Head Start for the same amount of money, or how many adults could get health care, or how many homes could be retrofitted to be energy efficient.

When you are ready, talk to your elected representatives about reducing the military budget to redirect that money to our social needs and true safety—including paying for climate solutions! As Ben Cohen said, "If one person does it, it doesn't have that much effect. If five do it, they start saying, 'Hmm, something's going on here.' If ten or twenty or fifty do it, and there's letters to the editors, too, then they say, 'Oh my God, what's going on here? I better pay attention.'" You can also urge your member of Congress to stop the massive cuts to the State Department so that we can respond to global conflicts with diplomacy, not violence.

Find out if you or any institutions you are affiliated with are invested in the military-industrial complex. Examine your investment portfolio, as well as your city's, your university's, your religious institution's, your bank's. The largest suppliers of the military are Lockheed Martin, Boeing, Northrop Grumman, and General Dynamics. Make sure nothing you're connected with invests in those companies.

The bottom line is only we as voters and organizers

can force our elected officials to change this unjust and unsustainable reality. With practice, you will become good at making people listen, and before you know it, more and more individuals and organizations will lift up their voices in unison with yours. And the benefits are enormous, including less war and unlocking practically unlimited funds to pay for the programs we urgently need to transition to a clean, safe, prosperous economy for all.

Jane looks down from the upper terrace of the Russell Senate Office Building. She had to step back during this week and could not risk arrest, but was cheering on the writers' room of Grace and Frankie and the other speakers, celebrities, and friends who were risking arrest.

CHAPTER SEVEN

Environmental Justice

I was away from D.C. the entire third week of November, which I found discombobulating. In New York, I attended **Glamour** magazine's annual Women of the Year event to receive that award on behalf of Greta Thunberg. After my weeks in D.C. embedded in climate activism and issues of environmental justice, I felt oddly out of place on the red carpet. This certainly wasn't my first time feeling discomfort from straddling the contradictions in my life, but I'd been slipping rather easily into the hand-on-jutted-hip stance of late, and I now wondered if the fish-out-of-water vibe was going to be my new normal.

Halfway into the evening, three young climate activists, dressed in their finest, spoke movingly about how Greta had inspired them, and then they introduced me. In my remarks, I emphasized how the young Swedish student, with her unique focus and way of cutting to the bedrock truth of the climate crisis, had challenged us to stop behaving as if this were business as usual, to listen to the scientists, to be bold, and to get out of our comfort zones. Then I read Greta's words, urgent and succinct:

> Unfortunately, I cannot be with you tonight.
>
> I'm incredibly honored to have received this award. And I'm very happy that it's been given to a climate activist—that would probably not have happened two years ago. Something has happened.
>
> If a Swedish teenage science nerd—who has shopstop, refuses to fly, and who's never worn makeup or been to a hairdresser—can be chosen a Woman of the Year by one of the biggest fashion magazines in the world, then I think almost nothing is impossible.
>
> That is hopeful, because that is what we need right now to prevent a climate catastrophe. We must do the impossible.

There was a rousing response from the audience. They had been moved, but would they be moved to act?

The Teach-In

I had to go from New York to Atlanta for the annual fund-raising gala of the Georgia Campaign for Adolescent Power and Potential, the nonprofit I founded twenty-five years ago. This event had been scheduled a year in advance for that Thursday, so I wasn't able to host the teach-in about environmental justice. June Diane Raphael, the actor who plays my daughter Brianna in **Grace and Frankie,** took my place alongside Annie Leonard.

Annie, an environmentalist for thirty years, talked about how in college she studied pollution, waste, and the environment, but when she graduated into the real world, she discovered that decisions involving who suffers the toxic consequences of those things occur in a racial, social, economic, gender, and power context that wasn't included in her environmental policy classes. If you want to promote climate and environmental solutions, you must take all of that into consideration.

She noted a report that had first opened her eyes to this, one that came out in 1987 called **Toxic Wastes and Race in the United States**. It looked at toxic waste sites all around the country to see what characteristics they shared. The report showed that the racial and class makeup of the community determined the location of where companies dumped toxic waste, not things like distance from an aquifer or permeability of the soil, as she'd been taught in school. The report documented clearly that communities of color and low-income and indigenous communities are more likely to host hazardous facilities. Hence, they are disproportionately polluted, disproportionately having health impacts, and, even worse, excluded from decision making about these facilities that are impacting their lives.

Annie was a hiker in her hometown of Seattle and thought about environment as being forests and animals. One day she heard an environmental justice leader named Dana Alston say, "The environment is where we live, work, and play." In other

June Diane Raphael co-hosted with Annie Leonard (at right) while Jane was out of town, and the guests included Von Hernandez, Yvette Arellano, Catherine Flowers, and Kerene Tayloe.

words, "environment" is everything that surrounds us through all parts of our lives, and that's a different way of thinking about it for a lot of people.

Dana and other environmental justice leaders challenged the big wildlife conservation and environmental groups to diversify and to include justice issues as part of their environmental campaigns. They said, "Your staff, your board, your members, don't represent our communities and don't demonstrate an understanding of them." We mustn't fool ourselves and think that we can win against pollution and climate without addressing the not-so-new reality of those living in the places where the problems are born. The poor and people of color are the first to feel the impacts of pollution because they "live and work" at the epicenters of extraction, refining, and exporting of fossil fuels. They are the ones breathing clouds of toxic dust or seeing brown flammable water coming out of their taps, day in and day out. They may be the first to feel the impacts of the pollution that leads to climate change, but those chemicals—carbon dioxide, methane, benzene, nitrous oxide, fluorine, and more—are altering the climate and the environment for everyone else. This is why to fight climate change we must also address the idea of environmental justice for all.

The Rally

I took a red-eye back to D.C. just in time for the Friday pre-rally briefing, which we held in a far bigger space because there were so many people that week. I was very moved and impressed that the entire writers' room from **Grace and Frankie** had come, along with our showrunners, Marta Kauffman and Howard Morris, a few of our producers, and June Diane and Brooklyn Decker, the actor who plays my other daughter in the show.

My workout trainer from Los Angeles, Malin Svensson, was there with her beau as well as my old friend the Vietnam veteran and California Democratic Party senior adviser Bob Mulholland. More than fifty people signed up for the civil disobedience and to risk arrest.

It was a boisterous, uplifting group of us that marched to the Capitol that morning, and we were met with a larger-than-usual crowd. The speakers who led off the rally were those who had witnessed the ravages of the lack of environmental justice first-hand. There was Kerene Tayloe, whose organization WE ACT took on the New York City transit, the MTA, because it located five of its six bus terminals in Harlem, greatly contributing to that neighborhood's 25 percent asthma rate in children, the highest in the city. They got the MTA to switch to buses that used less polluting fuel and to insulate the terminals more effectively from neighborhoods.

Kerene Tayloe speaks.

"Anything—you name it—that you don't want in your community, it's going to end up in black and brown communities," she said, summarizing the point of environmental justice. "That phrase, NIMBY, 'Not in My Back Yard,' well, when you have the political power to say, 'No. Not for me,' where do you think it's going to go? History has shown that it goes into black and brown communities. And we don't deserve that."

From Kerene, I learned about people in Alabama who are overwhelmed by coal ash because a corporation made some arrangements with the local government to dump the ash in their largely black community, exposing those people to cancer-causing particles in the air. I knew that white working-class people in Appalachia and in Cancer Alley in Louisiana were not exempt from toxic exposure. Reading **Strangers in Their Own Land** by Arlie Hochschild was beyond disturbing in its description of workers and families in and around that concentration of petrochemical plants and refineries who had cancers, asthma, respiratory illnesses, heart diseases, and rashes yet would not, could not, blame the companies because they feared losing jobs.

I had been to some of the places whose residents were in need of some environmental justice, although I didn't have the language to describe what it was at the time. I sit on the board of the Center for Rural Enterprise and Environmental Justice, and its founding director, Catherine Flowers, who spoke at our rally, is a friend of mine. She was born in Lowndes County, Alabama, and I went there with her one sweltering day for a board meeting.

Lowndes County is between Selma and Montgomery. It is flat and wooded and filled with civil rights history. Slavery might have been outlawed, voting rights might have been restored (although they are being rolled back in many places where it's evident that people would vote Democratic), but jobs are scarce, and it soon became clear that people lacked the right to have decent plumbing and access to solar heating. We drove down narrow dirt roads way back into the woods to visit trailer homes that had pipes carrying raw sewage straight into backyards where it pooled and festered. The families simply couldn't afford a septic system. Other homes were connected to the municipal sewage system, but the infrastructure was cheap and poorly maintained. When the rains got heavy, which is becoming more frequent, people had sewage flowing back into their homes. Often, they had to replace carpeting, even flooring, after the rains subsided. Catherine had been getting the word out to policy makers that hookworm, a Third World health problem, has been found in Lowndes County, and as we stood, stunned by what we were hearing and seeing (and smelling), we talked about how, with climate changing, there is the real threat of diseases that previously could not exist up north now coming here, carried in wastewater. Catherine had just read that bubonic plague has surfaced in the United States. "That's something from the dark ages," she exclaimed. "We're talking about these diseases coming back."

"Most elected officials don't venture down those back roads, so how can they begin to come up with solutions to problems they've never seen?" Catherine asked the crowd.

After Catherine, the next speaker was Abby Disney, a documentary filmmaker, philanthropist, and social activist. As the niece of Walt Disney and the daughter of Roy Disney, Abby grew up in a very wealthy, powerful, conservative family.

"I came to New York City in the mid-1980s, I grew up in a very pampered, fancy place, believe me, and I looked around me and I was like, 'What the hell is going on here? People are sleeping on the sidewalk, and other people are stepping over them to get to the

Abigail Disney speaks.

subway. What the hell is going on?' It was like falling into a capillary in a great big vascular system, and it led to a vein which led to a bigger vein that led to a bigger vein, and you know where those things lead to? They lead to your heart, because the work we're doing now around justice, the work we're doing for our earth right now, is about love. It's about radical love." And then, quoting Cornel West's famous line, Abby said, "Never forget that justice is what love looks like in public."

Abby got it just right. All these issues we have been focusing on are veins of a vast, interdependent vascular system. Some may feel their wealth and power will shield them from the growing crisis, as they have with the coronavirus, but the life-destroying effects of fossil fuels will, sooner or later, metastasize throughout the interconnected system. Sam Waterston was right when he said that in this situation empathy is enlightened self-interest. In some people, though, the empathy gene is muted. I pray that when these voices from the front lines are heard, those people will join the struggle, too, because they recognize that they will not be exempt.

The contrast between Abby and Yvette Arellano, a fiery Hispanic activist working with Texas Environmental Justice Advocacy Services, could not have been starker. Yvette grew up next to a toxic high school, César E. Chavez High in Houston, Texas, surrounded by a massive petrochemical infrastructure. She made the wry observation that "it's illegal for a petrochemical plant to locate itself next to a school but it's legal for a school to locate itself next to

Yvette Arellano speaks.

a petrochemical plant. These are the active loopholes that happen on a daily basis when new schools get built in communities of color."

Living all her life in this neighborhood, Yvette didn't know that other children weren't always surrounded by the putrid, sweet scent of benzene. She thought all little kids spent part of their summers locked indoors in 115-degree heat, windows covered by plastic, without air-conditioning, until the latest chemical disaster was over. The experience of "sheltering in place" was unknown to most Americans until the coronavirus but was very familiar to the many folks like Yvette. In her neighborhood, the fields were brown from toxins, and the children had to be warned not to play in the storm water because their parents couldn't guarantee that it was safe. When her high school football team played a home game, the fields were lit by the toxic flares of the petrochemical plants that surrounded them, young athletes breathing in those harsh chemicals. And the smoke that came from the chemical plants, children heard, was just water vapor. In fact, they called those toxic plumes cloud makers. This, she was told, was the price we had to pay for the convenient life we all lead. "But that's not true. Not everyone pays that price. And some people don't pay that price at all."

Fitting her personality, she asked not for pity but for people to take actions. She asked the audience to "remember the South. Please remember these communities. Remember us when we're not talking onstage with celebrities that are helping to uplift our voices. Thank you to all of you who brought us here, but the South can't be left behind. Don't let that happen. Come down and help us register voters and get the vote out. Help us in our fight against voter suppression. We need help."

It was surreal going from Yvette to Bobby Kennedy Jr., son of Robert F. Kennedy and nephew of President John F. Kennedy, but that is what I love about Fire Drill Fridays. We always include the voices of the most vulnerable communities along with the voices of those with economic and social power.

Bobby Kennedy, an activist and environmental attorney, has the gift of telling vivid stories that explain how and why power is made to yield. He suffers from spasmodic dysphonia, which makes his voice hoarse as if struggling to come out, and he can be hard to hear. I saw the crowd lean in to listen more intently as he told the story of suing the then mayor of New York, Rudy Giuliani. Maybe his vocal impediment was really a superpower like Greta's Asperger's, making people pay closer attention.

Bobby explained that New York City has some of the finest water in the world, the purest and the most delicious, flowing as it does directly from the Catskill Mountains. The water that comes into the city via the Delaware System is pure, clean, and clear. But the water from the Croton System on the East Side has water from the city's sewage treatment plants mixed in it. When Bobby asked for a map that showed which neighborhood got water from which system, the city refused. He had to sue to get it, and when he did, he saw the Delaware water went to the wealthy neighborhoods while the dirty Croton water went to Harlem, the South Bronx, the Lower East Side, and Hell's Kitchen, the poorest neighborhoods in the city. There was one little anomaly in the map, though.

"Over on the side, there was a little tiny white spot in the middle of the sea of Croton water, and there was one home getting Delaware water. I had to get a magnifying glass and look at it, and it was Gracie Mansion, the mayor's house. Rudy Giuliani was getting the good stuff."

The audience gasped and booed.

"This made it clear to me my advocacy as an environmentalist is about democracy, is about fairness, and is about equality and justice. These corporations are commodifying the commons," Kennedy said.

The concept of the commons is important for us to think about these days as we see more and more things being privatized. For our forefathers, the idea was we as citizens shared a common good, a common wealth. Although we defined it then as just for white citizens, the idea has survived to this day but in a much weaker form. Consider that the big park in Boston—the place where people used to graze their livestock communally centuries ago—is called the Boston Common because it is land that everyone in the community shared, like calling the state of Virginia a commonwealth.

Robert Kennedy Jr. speaks.

"Commons are the social safety net," Bobby said. "If you are poor, you don't have air-conditioning in New York City. You ought to be able to go down to the Hudson River or to Jones Beach and just swim. Those are the commons. During the Great Depression in the 1930s, ten thousand unemployed men in New York went down to the Hudson River and went fishing for food. Today, you couldn't do that, although the constitution of the State of New York says the fish belong to the people of the state. Today the fish are owned by the General Electric company. GE puts its PCP toxins in the river, and it's illegal to sell them anymore because they're so poisoned."

Bobby recalled his friend Lord David Puttnam, who in 2008 was advocating in Parliament for a cap-and-trade system to put a price on carbon. Although there was widespread support for cap and trade in Britain, members of Parliament said the country had to move incrementally so as not to cause economic disruption.

Lord Puttnam reminded Parliament that two hundred years earlier the debate about abolishing the slave trade had a similar context. Everyone agreed slavery was wrong and that it subverted the authority of Great Britain, which held itself up as a moral arbiter of the world. But slavery represented 25 percent of the British GDP. Free human labor was a huge source of wealth for the whole British Empire, and members of Parliament feared that if they abolished it quickly, the economy would collapse.

A year later, Parliament voted to abolish slavery, "and literally overnight and instead of collapsing, the British economy exploded as thousands of entrepreneurs rushed into that space to create new forms of energy," Kennedy said. "Well, today, we don't need to destroy carbon to understand that our deadly addiction to it is the principal drag on our prosperity and on American capital. According to the International Monetary Fund, the U.S. spends $649 billion subsidizing the coal, oil, and gas industries, more than the budget of the Pentagon.

"You don't need to have global warming to know that carbon is a bad idea. You just have to not like the acidification of the oceans. You have to not like the fact that we have cut down an area of Appalachia that's larger than the state of Delaware. And we have buried twelve thousand miles of streams in that state.

You just have to not like the fact that acid rain has sterilized 20 percent of the lakes in the Adirondacks and all of the high-altitude lakes in Appalachia from Georgia to northern Quebec. You have to not like the fact that we have mercury levels in every freshwater fish in America that make them dangerous to eat. You have to not like the fact that the fracking industry has poisoned the water in Dimock, Pennsylvania, and communities all over this country.

"It's about reclaiming our democracy. It's about ending the dirtiest, filthiest, most poisonous, most warmongering fuels from hell and moving to the cheap, clean, brave, wholesome, patriotic fuels we have. To do this means that you changing your politicians is more important than you changing your lightbulb."

I was shaken by the stories I had heard from our speakers but also elated about how diverse the lineup had been: frontline activists of color, Latinx climate leaders, a young climate striker, powerful, privileged (and older) white speakers who, nonetheless, had chosen to be on the right side of history. **This is what democracy looks like!** I have Annie and the Fire Drill Friday team to thank for much of the diversity. From day one, they had insisted that we never have an all-white or all-male lineup, and both the press and the movement had taken notice. A **Mashable** article said, "In the era of divisive generational burns like 'OK Boomer,' and increased youth-led protests and organizations, Fire Drill Fridays has created a safe, cool common ground where generations can work together."

WE'RE IN A DOCUMENTARY MOMENT

Mark Magaña

Mark Magaña is the founding director and CEO of GreenLatinos, a coalition of Latino environmental leaders.

The energy in the crowd was building when Mark Magaña took to the mic to inspire us all. He asked if anyone in the crowd had ever watched documentaries about a big movement like apartheid, women's right to vote, antislavery, farmworkers, or civil rights and asked themselves would they have been on the bus? Would they have been on the bridge? Would they have taken

Mark Magaña speaks.

the baton and the hose? Would they have gone on a hunger strike?

"You know what's fortunate right now? You don't need to wonder anymore. You are in a documentary moment. You are in a moment where your children and your grandchildren will say, 'Where were you? What did you do when we had a chance? Did you stand up? Did you fight back? Did you fight back? Were you on the ground?'

"You are in a documentary moment."

Catherine Flowers, Jane, and Ira Arlook (who ran Fire Drill Fridays' media work) march after the rally.

There were so many of us who marched to the Russell Senate Office Building that it took a long while to get us all through the metal detectors and into the center of the atrium to engage in civil disobedience by sitting down, holding our banner:

What do we want? Climate justice.
When do we want it? Now!

Bob Mulholland, my Vietnam vet friend, sat on the marble tile right behind me and I was so happy. For fifty years, Bob has always stood by me through all manner of dicey circumstances, providing physical security and moral support. Abby Disney said, "I was here when I was nine. My uncle Walt was being

Left to right: Malin Svensson (Jane's personal trainer), Bob Mulholland (behind Jane), June Diane Raphael, and Abigail Disney. They were occupying the Russell Senate Office Building.

presented the Presidential Medal of Freedom. He would croak if he could see me now," she said, laughing, very happy to be playing out her role of family rebel and true moral standard-bearer.

This time, when the police gave their third warning, I got up and went to stand on the balcony, which circled the entire rotunda, and looked down on the arrests happening below, loudly cheering each one as they were led off, most for the first time. I was very moved to see Marta Kauffman and the **Grace and Frankie** writers being led away, one by one, raising their handcuffed hands over their heads and chanting.

Because I had a pending court date and had been

held overnight previously, that was the first Friday that I was part of jail support rather than jailee, and I realized that my presence was just as much of a statement in that capacity. The fact that I had waited in the cold to hug each of them as they emerged from detention was meaningful. My view was, we're growing a movement, one hug at a time.

Because there were so many of them, it was well after dark before the last person was out.

When June Diane Raphael got out and arrived at jail support, she looked me straight in the eye and said, "Jane, this has been important," and I could see she meant it. She searched for words. "From this day, I will move forward differently in the world. Thank you."

Months later, when we were back at work, I asked her if she'd reflected on the experience, and she said, "I left Fire Drill Friday with two sometimes competing feelings. I realized that I need to make myself uncomfortable. Because I have the privilege of wealth and whiteness, I need to continue to put myself on the front line of activist movements on behalf of so many people who can't do so."

June went on, "Fire Drill Fridays and Greta's call to action has made me realize how addicted I've been to consumer culture. As the climate justice movement demands that we reckon with our destruction of the earth's resources, I've been reckoning with my own value system. I have been in the process of sifting through the ways in which I've upheld a capitalist system that does not value indigenous land, natural resources, and community. I am only on the beginning of this journey of realigning my time, energy, and talents so they reflect my values, but I'm feeling like I can breathe much more deeply."

I tried to tease out of our show writers if, by chance, because of their experience that day, Grace and Frankie might become climate activists in the final season, but I got no commitments.

What Can I Do?

Fortunately, there is a lot you can do to advance environmental justice, whether you happen to be a member of a disenfranchised community or from a more privileged and protected one.

Start by learning about and committing to incorporating an environmental justice approach at work—whatever that work is. The best place to start is the Jemez Principles, which were developed in the 1990s as a framework for just and inclusive organizing and collaborations:

Be Inclusive

Emphasis on Bottom-Up Organizing

Let People Speak for Themselves

Work Together in Solidarity and Mutuality

Build Just Relationships Among Ourselves

Commitment to Self-Transformation

Much has been written about the Jemez Principles, so I encourage you to google them to learn how organizations have incorporated them into their campaigns and cultures. Once you commit yourself to the Jemez Principles, any work you do on climate issues will advance environmental justice.

A good next step is finding out about the toxic exposures where you live, work, and play. Local organizations working on environmental justice may have information on neighborhoods near you, as well as campaigns you can join to advocate for stronger regulation and toxic waste cleanups.

If there isn't already a group working in your area, research the information yourself. Knowing which specific communities are exposed to which toxins, and in what quantities, is basic information that the public has a right to know and is a foundation for starting a campaign on environmental health and justice. The EPA no longer maintains an online map that shows which toxins are being released where. I recommend you contact your state or local environmental authorities to ask for this information about your area. You'll get mixed results, but asking questions of public employees helps pressure them to release this data, which should be tracked and freely available.

Sadly, but maybe not surprisingly, the environmental

movement has a troubling history of too often being racist and classist. For decades, many environmentalists assumed that people of color and poor people didn't care about the environment, and they hired outside "experts" to dictate solutions for local communities. Many times, they displaced indigenous people from their homes in an effort to "preserve" and "conserve" some so-called pure, untouched nature. Be sure to include those most affected by a problem in any proposed solutions, which results in better solutions overall. As environmental justice advocates demand, "Nothing about us without us."

Be certain that your campaign victories don't just push a problem elsewhere. For example, your campaign helps shut down an incinerator or coal-fired power plant only for the same polluter to move somewhere else. Our campaigns need to move beyond NIMBY (Not in My Back Yard) to NOPE (Not on Planet Earth) so that when we stop a fossil fuel project or other polluter, it's gone—not just moved.

Remember, we are **all** affected by environmental problems one way or another. Those of us with more power and greater influence need to collaborate with those on the front lines, rather than condescendingly "helping others." That's how we build just relationships for just futures. Think solidarity, not savior.

There are thousands of local and regional groups working to advance environmental justice, so this is a way-too-small list of suggestions of efforts to support or join.

In the southern United States, check out Texas Environmental Justice Advocacy Services (T.e.j.a.s.), the Deep South Center for Environmental Justice, and the Alabama Center for Rural Enterprise Community Development Corporation (ACRE).

For environmental justice campaigns led by indigenous peoples, check out Honor the Earth, the Indigenous Environmental Network, and, in Canada, Indigenous Climate Action.

In New York, check out WE ACT for Environmental Justice (WE ACT), NY Renews, and Green Worker Cooperatives.

In California, check out Communities for a Better Environment; the Center on Race, Poverty, and the Environment; STAND-L.A.; and Movement Generation.

And, of course, there are many climate justice groups like Climate Justice Alliance and many longtime environmental groups like Greenpeace, Friends of the Earth, and the Sierra Club that embrace environmental justice principles in their work.

Jane stands onstage with Maude Barlow,
one of the speakers.

CHAPTER EIGHT

Water and Climate Change

Senator Ed Markey and his wife, Susan Blumenthal, were friends of mine from the 1990s. At the time, Susan was assistant surgeon general. Her specialty was women and suicide, and I had sought her expertise when I needed to understand my mother's suicide while writing my memoirs. I was delighted when Susan reached out to me on behalf of her husband, the senator, and said he wanted to invite me to speak with the Senate Task Force on Climate Change. Senator Markey is, after all, the co-sponsor with Alexandria Ocasio-Cortez of the Green New Deal.

Annie Leonard and Ira Arlook were with me when I went to a meeting room in the Hart Senate Office Building. Staffers for the senators on the Climate Task Force sat in chairs lining the walls, and I had learned from my lobbying experiences back in the 1970s that sometimes it can be more useful to plant ideas in the minds of smart staffers who can, in turn, influence their boss's legislation.

I had been forewarned that senators would be coming and going because there were important votes taking place that they needed to be present for. But Senators Markey (D-Mass.), Kirsten Gillibrand (D-N.Y.), Jeanne Shaheen (D-N.H.), Tom Udall (D-N.M.), and Jeff Merkley (D-Ore.) were there for most of the meeting. I explained why I had started Fire Drill Fridays, but mostly I wanted to know what they thought and what they were willing to do. "Am I doing what I should be doing?" I asked them. "Do you think there's a better way to rouse people to the urgency of this crisis?"

It was Ed Markey who answered first, saying emphatically, "Yes! You're building an army. Make it big. We need that public pressure from the outside."

Given the current administration's hostility to anything related to slowing climate change and preventing water and air pollution, I could understand the enthusiasm with which the senators responded. They needed all the help they could get, and they surely know that no big transitions in history have ever happened without an organized, angry public demanding them.

A key question Annie and I wanted to ask was, were the senators willing to push forward climate legislation with a majority of senators, even if they don't have sixty votes? When they resoundingly answered yes, we cheered. This was important to know because it is not likely we'll get sixty climate leaders serving as senators for years, and failure to reach that super-majority had stymied progress on climate and many other important issues.

On a less positive note, when I asked them what they thought about Fire Drill Fridays' three goals, they all agreed with supporting the Green New Deal, but when it came to halting fossil fuel extraction and starting a just transition off fossil fuels, instead of addressing that issue, they immediately pivoted to how much they all support 100 percent renewable energy. That's great. I support 100 percent renewable energy, and I am glad they do, too. But I explained to them that if they continue to permit fossil fuel expansion, it will cancel out much of the benefits of the increased renewable energy. As the renewable energy and electric vehicles reduce demand for fossil fuels in the United States, the newly drilled fossil fuels will be either turned into plastics or exported—either way, continuing to drive the climate crisis.

I was surprised that they were far less willing to take this stand. I find it disturbing that even friendly elected officials who somewhat understand climate and support a Green New Deal aren't ready to say we need to stop expanding fossil fuels. Clearly the climate movement must keep demanding it, reminding them of the science. If the IPCC says we have to drop fossil fuel emissions by 50 percent in ten years, why on earth are we continuing to expand them?

But believing you must celebrate what you get

and keep fighting for the rest, Annie, Ira, and I left the building feeling pretty great. On the way to the elevator, we ran into Mark Ruffalo, who was meeting with senators to encourage them to take action on a very toxic class of chemicals found in many household products, called PFAS, that are the subject of his gripping film, **Dark Waters**. Just like the case with fossil fuel corporations, the companies that produce these chemicals have known about their serious negative health impacts for decades and have invested in lobbyists to block regulation and in public relations to promote a positive image, even while continuing to knowingly pollute communities. This is the exact same issue as with fossil fuel corporations. It is systemic. We can't win by fighting one chemical at a time or one pipeline at a time. To reclaim our democracy, we must rid our government of the corporate stranglehold and demand that our elected leaders prioritize public health and environmental sustainability over the interests of corporations that have proven they are willing to sacrifice both for profit.

It was a relief, the week leading up to our seventh Fire Drill Friday, to not have to travel, to stay home—yes, my hotel room had become home—and study. I was being sent climate-related articles and books by the truckload, and it was hard to keep up.

That Fire Drill Friday was focused on water. "**Water is life. Mni Wiconi**." I first heard that statement in 2016 on the Standing Rock Reservation, where the water protectors fought the heavily militarized police to save their water from the Dakota Access Pipeline. Without water, there's no life. This was abundantly clear at Standing Rock, where the proposed route of the pipeline crossed underneath the Mississippi and Missouri rivers, threatening drinking water and water used to irrigate farmlands. Through the people I met and the studies I read about water in the weeks leading up to our Fire Drill Friday, I learned about the precarious state of clean water for everyone worldwide. Already, billions of people globally lack access to safe and clean water, and climate change is going to make this worse.

For most Americans, this is unimaginable. We're taught in school about the water cycle, the idea that water is infinite and endlessly replenishing. The heat of the sun draws the water up into the clouds, which rain it down on our cities, mountains, and plains. It travels down the hillsides and through the gutters into the streams and our water systems, until the municipal water plants extract its impurities. When we turn on the tap, we do not doubt that what comes out is abundant and clean.

Water is no longer infinite. With industrial and agricultural pollution, as well as fracking, we're using it up, and the consequences of this are now visible and accelerating. The UN predicts that by 2050 that lack of access to clean water could affect five billion people, half the planet. We may feel safe here in America, but when the U.S. Government Accountability Office surveyed water managers in 2013, it found that forty out of fifty of them anticipated water shortages by 2023. Yet few talk about this coming scarcity, which is directly related to, and made worse by, climate change.

The Teach-In

I had learned so much about our water crisis, and how it was interlinked with the climate crisis, by reading an excellent book, **Whose Water Is It, Anyway?**, by Maude Barlow, one of the world's leading water experts.

Annie Leonard invited Maude to our water Fire Drill Friday, telling me I would adore Maude, and she was right. The slight, gray-haired woman was probably the number one water advocate in the world. Maude is the national chairperson of the Council of Canadians, chairs the Food & Water Watch board, co-founded the Blue Planet Project, which works internationally for the human right to water, has served as senior adviser on water to the sixty-third president of the UN General Assembly, and has written nineteen books on water. She is also warm and generous. I right away felt we would become friends.

It was appropriate that Maude was our first speaker at the teach-in because she saw all of this coming. In the 1980s in Canada, she noticed that businesses were trying to buy up water rights to privatize what should be a public good, a resource grab that has become more aggressive in the last three decades as water has become scarce internationally. Once a private company takes over a municipal water supply, the rates increase by as much as 59 percent, becoming unaffordable for many. Some municipal water companies have the right to turn off your family's water or even seize your home if you fall behind on your water bill, which has started to happen in cities around the country. Even during the COVID-19 crisis people's water is being turned off in some cities!

The first thing Maude said at the teach-in was that she wanted to elevate the water issue in the climate debate so that instead of seeing water as a subset of climate change, we come to understand that abusing water is actually a major contributor to climate change.

"In fact, I believe that if we were able to stop every greenhouse gas emission in the world tomorrow, we would still have a water crisis. And that's because we have polluted and diverted and dammed our rivers to death. We're extracting groundwater way faster than it can be replenished by nature. We've created an ecological crisis by moving water to where we want it from where nature put it. And so it's really important for us to understand that not only is it a major cause of climate chaos, but the reverse is also true. That protecting and promoting the well-being of watersheds is a major solution to climate change. It is also the major human rights issue of our time in terms of the sheer numbers. More children die of waterborne disease than all forms of violence put together, including war. So there's an ecological crisis and a justice crisis."

I thought about how many people who have spoken at Fire Drill Fridays have equated climate and justice.

"Because we're a planet running out of water, there are people who are trying to control it, privatize it, commodify it. To make money from it. Because they knew before many of us that whoever controls water will be both powerful and wealthy.

"And there are literally life-and-death decisions being made today, right now, by governments and leaders, about who gets water and who doesn't. And it usually isn't going to the villages who need it. No, it goes to mining companies, fracking companies, water-bottling companies. The United Nations has declared that access to clean water is a human right. And it's a public trust and we must maintain democratic control over what happens to water. We're not waiting for our federal governments; we're trying to bring this concept of the human right to water into local communities through municipalities, through faith-based institutions, through educational institutions. We have to do this at a community level because we're in a race against time, Jane. The water crisis is here. It's now. And it's not just in the global south. It's right here in the global north as well."

Again, a crisis of the commons.

Mary Grant was the other guest at the teach-in. She is the Public Water for All campaign director at

Food & Water Watch, overseeing campaigns to support universal access to safe water in the United States by promoting responsible and affordable public provision of water and sewer service. Her focus was not just on access but on the aging water system, how our water infrastructure is outdated and not built to handle the floods and droughts that will come with climate change in the twenty-first century.

"What we're seeing is intense and extreme drought, which means water shortages across the country, particularly in the Southwest and the southern Great Plains, leading to overuse and depletion of groundwater. At the same time, we're also getting these intense heavy rainfall events, particularly in the Northeast and the Midwest. And this is overloading and flooding our wastewater systems, leading to sewage spills that are polluting our waterways."

Mary highlighted another threat to our abundant

Mary Grant speaks at the rally, as Jane and Maude Barlow watch. Mary is holding a bottle of tap water from a home in Pennsylvania near a fracking site.

freshwater: toxic algae blooms in our Great Lakes and in lakes in Florida. Algae is naturally occurring, but when the lakes receive high levels of nitrogen from fertilizers from factory farm runoff or effluent from leaky septic tanks, this causes the algae to grow wildly and give off toxins. If it grows without check, it can literally poison our Great Lakes and other bodies of water.

"And the climate crisis is creating the perfect conditions for it," Mary said. "We have rising global temperatures, warming waters, and at the same time more pollutants in the water because of heavy storm-water runoff. It's a perfect breeding ground for toxic algal blooms. If we don't act now, more and more communities will have toxic water."

Protecting our water used to be a national priority, but the federal government has cut funding for water by 74 percent since 1977. Mary is part of a coalition advocating for Congress to pass the Water Affordability, Transparency, Equity, and Reliability Act (the WATER Act) to address the fact that fourteen million households are struggling to pay for water that is not even safe to drink. It will take federal money to solve this problem, and as with many of the infrastructure needs discussed in the Green New Deal, doing so will create jobs for many, Mary noted.

"Nina McCoy is an activist for Martin County Concerned Citizens in eastern Kentucky, one of the most impoverished counties in our country that has a real water emergency right now. And she posed this question to me: 'How do we get these out-of-work coal miners into jobs fixing water pipes?' Bringing workers who work in these dirty fossil fuel industry jobs that're polluting our water, and bringing them into the water sector, to fix our pipes, to build out systems to communities that lack it."

My last guest at the teach-in, Jessica Loya, with GreenLatinos, a national network of Latino environmental advocates, joined in. "We have seen water contamination from fracking, we have seen water contamination from big agriculture, we have seen water contamination from oil spills. The federal investment in water infrastructure has gone from sixty-three percent of federal agency spending in 1977 to less than nine percent today. There are communities in south Texas who have never had running water or sewage management systems. The infrastructure just isn't there. And we see the water problem internationally. El Salvador, Guatemala, and Honduras (known collectively as the Northern Triangle) are experiencing a decades-long drought, impacting indigenous communities' ability to have subsistence farming, requiring them to go to cities, and then leading them on a dangerous journey across borders to reach the United States, only to be turned away. Water knows no borders. People are risking their lives for water today, and it will get worse."

I wanted to know about the role desalinization of

seawater can play. Is that a way for us to address water shortages?

Mary answered, "No! It's energy intensive, and it pollutes because it takes in the ocean water and all the aquatic life, separates out the water, and puts the salt brine with the chemicals and the dead aquatic life back into the ocean. It is really not a long-term solution, because it ends up killing the oceans."

Maude added, "There are two cities in southern Australia. During the water crisis a couple of years ago, one went heavy desalination, and the other built hundreds of wetlands and put in the kind of plants that eat pollution, and they created a miracle, they've created an oasis in the desert, just doing it with nature as opposed to all the technology that we use that says that we're somehow superior to nature."

Mary had a message of hope, too. In Baltimore, where she's from, people were losing their homes because they couldn't pay their water bills. Voters there took on the water issue themselves. "We just held our elected officials accountable. We grew public power, public pressure. Baltimore voted to ban water privatization. We declared our water and sewer systems inalienable assets of our city. And then our city council passed an ordinance to set up an income-based water affordability program. This is a comprehensive solution to make sure that every person in Baltimore can afford to pay their water bill."

Mary's Baltimore story lifted my spirits as all stories of people regaining their power do.

KEEP WATER PUBLIC:
GO BLUE

Maude Barlow

"We call them Blue Communities, and we started it in my country, Canada, ten years ago because we had a government that refused to recognize the human right to water and was promoting privatization of water services. So instead of being against that, we decided we wanted to be for something, so we came up with this thing we called the Blue Communities Project. And basically, it was a way to fulfill at a local level this commitment to the human right to water. And we thought of it at first only as municipalities and frankly only in Canada. To become a Blue Community, a municipality has to promise to protect and promote water as a human right. To protect and promote water as a public trust, to keep it in public hands because you don't want private corporations deciding who gets access to it because it's going to be on a for-profit basis. And the third starts to address the plastics issue, which we haven't talked about yet. But every single moment, in our world, a million plastic bottles of water are sold. If you put them together end to end, every single year they would go past halfway to the sun. I mean we're just destroying our water systems with this plastic. So the third commitment of the municipality is to do away with plastic bottled water on municipal premises and events and actually to promote safe drinking, accessible drinking, water within the community. And it took off and we have something like, I think, thirty-five municipalities in Canada, from the small indigenous ones to Montreal, which just became our latest Blue Community. But then it spread to Europe, and we have big cities like Berlin and Munich. Paris became a Blue Community when they kicked their private water company out. They started a public company called Eau de Paris, which I think is absolutely perfect."

The momentum is starting to go in the direction of cities taking back their water supplies and other public utilities. When cities do this, the benefits are immediate. In the case of Eau de Paris, the city saved $39 million in the first year, slashed prices by 8 percent, and used the money it saved to reinvest in maintenance rather than paying shareholders the profits. Paris even installed free sparkling-water drinking fountains around the city.

Paris was just part of a re-municipalization movement that has been sweeping the world in the last decade. A 2019 study of this trend called "The Future Is Public" found that between 2000 and 2019, 1,408 cities worldwide returned privatized public services like water, telecommunications, energy, and health care to public control. That number included 311 water and sewage utilities brought back into public hands, with 71 of those in the United States.

The report described how privatizing public services and public-private partnerships was losing credibility with citizens who suffered price hikes and reduced quality of service under private control. Publicly owned companies devote their services to the public good, rather than trying to maximize profits. As publicly owned municipal utilities lower costs and extend services to underserved neighborhoods and support local businesses, they forge a new sense of the public good, increasing public participation in decision making. And, with benefits like free sparkling water on tap, they foster an increased sense of public pride, as Maude found with her Blue Communities.

"So now we have universities becoming

Blue Communities and doing away with bottled water. The World Council of Churches, 500 million Christians, have become a Blue Community and are spreading the word around. And we met with a public school in my city where two of my grandkids go, and we met with the teachers last week, and they're going to become our first elementary Blue Community school. We want the kids to get involved because that's where it starts. If the kids go home and say to their parents, 'Why are you buying bottled water? Do you know that that plastic bottle of water contains way more microplastics than your tap water? Do you know that, Mom and Dad?' If we start with that kind of education, we can really move mountains."

The Rally

We had many celebrities for that Fire Drill Friday: My pal Diane Lane and Manny Jacinto, the young actor who plays the pretend Buddhist monk on **The Good Place,** had come, inspired by his castmate on the show Ted Danson, who had risked arrest several weeks earlier. The actor Piper Perabo and the supermodel Amber Valletta were also there, all of them willing to brave the pouring rain.

I hadn't wanted a focus on water without having someone from Standing Rock be represented.

I introduced Alice Brown Otter, a sixteen-year-old water protector and the second indigenous youth to attend Fire Drill Fridays who had made the 1,519-mile run from Standing Rock to the steps of the Army Corps of Engineers with a petition to stop the pipeline.

"When I was twelve, I really didn't know any of the logistics or anything of pipelines," Alice told the crowd. "I kept asking myself, why are people so selfish? Why isn't my life as important as everyone else's? The thing that I thought I could do was run. So I ran all the way from North Dakota to here. Now, four years later, I'm here.

"The thing that I have to say is, why are we waiting? How long is it going to take for everyone to stand up and to realize how important water is and how much it sustains us? How come on social media I just see stupid things that no one should be caring about, but

Alice Brown Otter speaks.

I don't see anything about climate change? We gotta wake up!"

Piper Perabo urged everyone to reach out to their representatives and tell them that they want them to support laws that protect our groundwater.

"And while you've got them on the phone, encourage them to restore the federal investment in America's water and sewer infrastructure because there are more than two million Americans that lack running water and access to sanitation," Piper said. "And the WATER Act will help fund our country's water infrastructure, which we sorely need. If you don't have your phone out already, get out your phone, and type in the number; I'm going to give you the number for the U.S. House switchboard operator so that this afternoon you can call your members of Congress. Ready? The number is 202-224-3121. Save it to your favorites. Call your member of Congress and ask them to support the laws that protect our water." I was impressed that Piper had come prepared with that number to give people. It's a number we should all program into our phones and call daily to urge our elected representatives to act on the issues we care about.

Next, Mark Magaña of GreenLatinos described his group fighting a massive pipeline that is going to be laid between the United States and Mexico to transport natural gas from the Permian Basin, the largest deposit of hydrocarbons in our nation in an area about the size of the state of Oklahoma. Imagine the size of Oklahoma full of communities that can't breathe because of 24/7 fracking operations that are extending from West Texas to North Texas to New Mexico.

"All of them basically are sitting on top of our country's largest freshwater aquifer, the Ogallala Aquifer, that extends from Texas all the way up through Oklahoma into Nebraska. And so we're also dealing with the fight for freshwater," Magaña said. "Folks don't realize that communities along the Gulf Coast and border communities, both incorporated and unincorporated, stand alone in trying to sustain themselves off of water wells that have been contaminated by these 24/7 fracking operations."

Manny Jacinto, born in the Philippines and raised in

Canada, stood under an umbrella I was holding over him, rain still dripping off his fan-favorite jawline.

He told us, "According to the Global Peace Index in 2019, the Philippines is the most susceptible country to hazards brought about by climate change, and at least ten million Filipinos have no reliable access to safe and improved water sources.

"The climate emergency is already aggravating the water situation with droughts, tsunamis, and uncontrollable floods. And as we stand here, another tropical storm is dumping massive amounts of precipitation in the northern Philippines.

"While the Filipinos are at the forefront of the global climate emergency, they're also leading the fight for climate justice and solutions just like all of you here." Just a few weeks later, on December 9, the Philippine Commission on Human Rights found,

Jane holds the umbrella for Manny Jacinto as he speaks.

after a three-year investigation, that the biggest fossil fuel companies, the forty-seven carbon majors like Exxon and British Petroleum, played a clear role in climate change and were legally liable for its impact. Although the decision carried no legal weight, it could be used to strengthen laws to hold the carbon majors accountable for damages to the country.

I then stepped up to hold the umbrella over Garett Reppenhagen, an Iraq veteran, son of a Vietnam veteran, grandson of two World War II veterans. He served in the U.S. Army as a cavalry scout sniper in the First Infantry Division. Garett gained an honorable discharge in 2005 and began working as a veterans' advocate. He served as the chairman of the board of Iraq Veterans Against the War and as the vice president of public relations for Veterans for America, and as a program director for Veterans Green Jobs. He lives in Colorado, where he serves as the executive

Garett Reppenhagen speaks. Amber Valletta and Manny Jacinto stand behind him onstage.

Amber Valletta, Piper Perabo, and Diane Lane get arrested as the group occupied the street in front of the Capitol.

director of Veterans for Peace and is organizing veterans to demand climate solutions.

"You know I live in Colorado, and in the West we have a saying that whiskey is for drinking and water is for fighting. After I joined the military one month before September 11, I've done a lot of drinking of whiskey. I've done a lot of fighting. And most of that fighting was for the exploitation of other countries, mostly their natural resources, protecting the extraction and transportation of fossil fuels. They have destroyed communities around the world. I may have fought for oil in Iraq, but the next wars are going to be fought over water. The lack of resources of water is going to make the next generation of Veterans for Peace members veterans of the Great Water Wars. And wars are generally destructive to water. In Iraq, I saw the demolition of an infrastructure that helps get people clean water, destroying the infrastructure of waste systems that contaminated water systems, the pollution and contamination of chemicals and weapons into groundwater, into rivers, that have destroyed communities. I fought in the Fertile Crescent that was nurtured by the Tigris and Euphrates river that is now too polluted to even swim in. Today, precious few can count on something so simple as just a cup of clean drinking water. Veterans for Peace has an Iraq water project where we're helping to deliver to communities, with the assistance of some great organizations like Life for Relief and Development to get purification units, ultraviolet units, reverse osmosis units, to communities in Iraq so they have clean drinking water. Most of these communities are hospitals, clinics, and schools. Veterans also made up a large contingent of water protectors in Standing Rock."

Because of my pending court date, I had to avoid arrest again this week. By the time the arrestees started to emerge from warehouse detention, the rain had all but ceased, which made it easier to set up jail support. Because the numbers of people willing to risk arrest had kept growing, we'd been asked to move to a sloping area across the street, which meant we could see the people as they stepped outside and could start cheering as they walked the hundred yards to where we waited. This week there were folks from Washington State, Oregon, Colorado, and Wisconsin, among others, and I was stunned. They had read about what we were doing and decided to come. Many had never done such a thing.

What Can I Do?

As our teach-in and rally guests explained, water really is life; it is necessary and irreplaceable and deserves ambitious programs to protect it for all. There are many things we can do to protect water for health, for climate, and for justice.

A good first step is to learn more and start conversations with friends, neighbors, classmates, and co-workers. Many people don't realize the threats to water. Some good resources are any book by Maude Barlow, starting with her latest, **Whose Water Is It, Anyway?,** and the short animated film **The Story of Bottled Water**, which is available online. Your municipal water system is required to test water regularly and share these reports with the public; request the latest annual report on your local water quality. If the report reveals problems, use it to insist on changes.

Other things you can do at an individual or household level are commit to not buying bottled water, carry your own refillable water bottle, and avoid hazardous household cleaners and garden pesticides that contaminate our waterways. If your home has a lawn, replace it with native plants that don't require watering and, if you live near wetlands, advocate for their protection because they are invaluable for water health. Each of these provides opportunities to act in alignment with our values, but they are just a drop in the proverbial bucket. To make bigger change, we have to work collectively on bigger goals.

Water is not something that can be protected by individuals alone or left to the market to solve. Start a campaign to get your city—or your university, your church, or any community—to become a Blue Community. The Blue Communities Project provides you with the tools to promote the human right to water. This project works with local governments, community activists, and water operators to ensure water justice for all. As Maude explained, the Blue Communities Project framework calls for the following:

- **recognizing water as a human right**

- **promoting publicly financed, owned, and operated water and wastewater services**

- **banning the sale of bottled water in public facilities and at municipal events**

You can get more information by searching online for the Blue Communities Project.

And while you're working to become a Blue Community, you can join with friends and declare your office, your church, your gym, your school, bottled-water-free. Engaged citizens in some cities like San Francisco have persuaded their governments to prohibit bottled water sales on city-owned property and at the San Francisco airport. SFO used to sell more than ten thousand plastic water bottles each day, almost four million plastic bottles a year! The airport installed a hundred water stations where passengers fill their own bottles, saving money and avoiding tons of plastic waste. Water stations pay for themselves through the savings achieved by not having to handle so much plastic waste.

On the national level, the Water Affordability, Transparency, Equity, and Reliability Act, which Mary Grant told us about, is really important. The WATER Act is a comprehensive approach to ensuring that every person in the United States has access to safe, clean water. To achieve our goal of safe, healthy water for all, which helps protect public health and combat climate crisis, we need a major federal investment to renovate the old, lead-ridden pipes in our water infrastructure and help towns that are affected by PFAS contamination. The WATER Act will simultaneously deliver water justice to millions of people in the United States by stopping sewage overflows, averting a looming water affordability crisis, and creating nearly a million jobs. You can find out more about the WATER Act from Food & Water Watch, including tips on contacting your Congress members to urge their support.

We can also ensure that our cities stop turning off the water when residents can't pay the water bill. Depriving someone of water is a health and human rights abuse; people in need should be supported, not made to suffer.

The group takes to the Capitol steps after the rally.
Pictured with Jane are Lindsey Allen, Iain Armitage, Paul
Scheer, Jane's grandchildren Viva Vadim and Malcolm
Vadim, Jane's step-daughter Nathalie Vadim, Ricardo
Salvador, and Jim Goodman.

CHAPTER NINE

Plastics

If I could have spent one more week in Washington, D.C., that Fire Drill Friday would have focused on plastics. But even without a plastics-focused week, the topic came up in many teach-ins just as it's showing up in all the world's waterways, in sea life, and even in our own bodies. And of course it is integrally connected to fossil fuels. This is why I want to include a chapter on this immense and climate-related problem.

I was born in 1937, at least twenty or more years before plastic was common in American households. School lunch was wrapped in waxed paper. Milk came in glass bottles that were returned when empty. Same with Coke. In **Life** magazine in 1955, there was an article, "Throwaway Living," about an American family celebrating the convenience of disposable plastics.

I remember vividly around 1963 in the airport when I was on my way to Geneva, Switzerland, with my

French husband, a Swiss banker tried to persuade me to invest in oil. "They can make clothes from oil now! Coats, shoes, you can't believe what they can make out of oil." I found it a questionable concept at the time, and it certainly would have been unthinkable, even then, that before long most people would find it hard to imagine life without plastic and the other things fossil fuel can be refined and concocted into.

But along the way, plastic went from being a convenience that was celebrated to a very serious environmental and health threat and a significant driver of fossil fuel extraction and climate change. Plastic production and plastic waste have grown exponentially during my lifetime, and now we, along with millions of other animals, are choking to death on it.

Von Hernandez, an award-winning environmental activist based in the Philippines, spoke at one of our Fire Drill Fridays about the interconnected effects of the climate crisis and plastic waste in his country. Although it is a small country, the Philippines ranks third globally in plastic pollution with masses of plastic trash clogging its waterways and canals and thick on its beaches, and he expects the problem to get worse. As renewable energy increasingly displaces fossil fuels, he said, the big oil companies see single-use plastic as their next big profit stream.

"Let me tell you an example where the fossil fuel industry has been pulling a con job on the planet, on the people of this planet," Von said. "It's not just the climate crisis that they have foisted upon us but also the plastic pollution crisis. Fossil fuel companies want to retain their profit margins by hedging their bets on more plastic production. And at the moment the petrochemical buildup and expansion we are seeing here in the United States would trap us into even more carbon emissions in the future. If current trends continue, it is estimated that in a few decades 20 percent of all the oil that will be produced will be devoted to plastics."

British Petroleum, Exxon, and Chevron estimate that they might make as much as 70 percent of future revenues from plastics, meaning plastic production alone will account for 15 percent of the remaining carbon budget that we have to burn from now until 2050.

Via his work with Break Free from Plastic, Von is coordinating a global network of organizations to fight plastic pollution at the source, the fossil fuel companies including fracking sites, all the way to the end point of incineration and landfills, with the goal of zero waste and proper recycling. As Denise Patel, his colleague from Break Free, pointed out, plastic generates toxins from the moment fossil fuel companies extract the raw material from the ground, all the way to when we use it in our homes.

"The plastics that you see, the plastic bottles, plastic film, plastic straws, the single-use plastics all start out at a fracking site or an oil well," Denise said. The by-products of manufacturing plastics are toxic, and up to 77 percent of the plastic we use at home gives off some toxins, according to a study published in 2019. The transportation of oil to the factory where it is made into plastic is toxic, leaking chemicals into the air and water that cause asthma, heart disease, and cancers, thereby adding to pollution long before it becomes plastic waste. Denise emphasized that we cannot solve the two problems separately but must fight starting at the extraction point all the way to how we dispose of plastics.

Denise explained that many U.S. cities now have recycling programs to collect plastic packaging that we use once and throw away. However, there isn't a market for the collected plastic in the United States, because it is cheaper to make new plastic directly from fossil fuels (called virgin plastic) than to collect and process discarded plastic waste. This is even more true since the fracking boom of the last decade made natural gas—a feedstock for plastic—so abundant.

"So rather than reduce plastic packaging at source and develop a workable recycling system within the U.S., cities here sold collected plastic waste to China and other countries in Southeast Asia, where it was recycled in operations that would never have been permitted in the U.S. But last year, mainland China, Malaysia, Thailand, and other countries that we used to ship our waste to started rejecting this imported plastic. The amount of plastic waste people in the U.S. were discarding was too much for these countries to manage, and communities living in places where the imported waste was ending up started pushing back, concerned about the amount of trash being dumped and finally clogging their rivers and polluting their oceans.

"Part of today's waste crisis here in the United States is that our cities and communities are not able to sell the plastic collected for recycling anymore and increasing amounts are going to landfills. Even worse, a lot of them are going to incinerators. Burning plastic emits greenhouse gases and dangerous, cancer-causing toxins like dioxins.

"Break Free from Plastic has done beach waste audits around the world to collect and identify the companies actually responsible for the plastic packaging in the ocean. They found that the number one company that is putting plastic into our environment is Coca-Cola, followed by Nestlé, Pepsi, and Unilever. So one of the things we can do to solve this problem is push those companies to use refillable or reusable containers rather than single-use plastic packaging. We also can mobilize to get local communities and cities to adopt goals of 'Zero Waste,' which is a comprehensive approach to solving waste at the source, designing it out of the system rather than trying to contain it once it is made and released into the environment. Yes, we should be responsible in our personal choices, but also governments and corporations have a huge role in solving the waste problem, which would help protect our oceans, public health, and the climate. We've seen citizens, when they're organized, forcing governments to ban plastic bags, plastic straws, and disposable plastic food ware, which is a good start since those are items frequently found in ocean plastic pollution."

It's true: All over the world, people are organizing to place restrictions on disposable plastic, and it is working! In the United States, more than four hundred cities and a handful of states have banned or taxed disposable plastic bags to reduce their use.

Annie told me about the ambitious plastic reduction efforts in her hometown of Berkeley, California. Not only did they ban most disposable plastic bags, but they worked with local restaurants and anti-plastic activists to develop a comprehensive program to eliminate a wide range of disposable plastic food ware—the take-out containers, cups, and cutlery that are a big part of the trash littering the city and making its way to the ocean. Berkeley's new program is even piloting a reusable take-out cup system; customers can "borrow" a stainless-steel cup for to-go coffee and return it later, just as we used to do with videos and DVDs.

Denise continued, "Fifty percent of all the plastic waste that's ever been produced was produced in the last fifteen years. And a lot of that is because of the fracking boom that happened across the United States. And only 9 percent of plastic that's ever been produced has been recycled." That means that more than 90 percent of all plastic ever produced has ended up being dumped, burned, or buried. We've been told that recycling is the solution to plastic waste, but most plastic is not recycled, because either it is not technically feasible, it's not economical, or there's no recycling infrastructure to manage it.

"We need systemic change here, and it starts by holding plastic and fossil fuel corporations accountable for the impacts of their harmful products and practices. We want to make sure we're voting for the people who are really going to make that happen. It's good to make lifestyle changes that reduce our carbon footprint because it's very important to practice our values. But we also have to take the next step and join the political fight because the fossil fuel industry is not going down without a fight. We're in the middle of it."

What Can I Do?

There are many ways we can all help reduce plastic. At home, avoid unnecessary packaging and use reusable shopping bags, water bottles, and even take-out containers. Buy products in bulk, which reduces packaging, and buy fresh food that has little or no packaging. When available, choose paper, glass, or metal over plastic goods. That's all good and can help us feel better about how we manage our households, but to have a positive impact at scale, we need to focus on much bigger targets: corporations and our government.

Alone, we don't have influence over big corporations,

but together we can hold them accountable for how they package their goods. Break Free from Plastic can give you tools to do a brand audit in your next beach cleanup, determining which corporations contribute the most plastic pollution so you can then pressure them to change. Greenpeace is running campaigns pressuring large retailers—like Trader Joe's, Target, and Safeway—to reduce the plastic packaging in their stores, and there are lots of ways to get involved. As John Hocevar, Greenpeace's oceans campaign director, explained during our oceans and climate teach-in, every time we go to our supermarket, whether it's a comment card or talking to the supermarket manager, we should let them know we need them to do better on plastic: get rid of plastic bags and rethink their plastic packaging.

The most effective way to reduce plastic in our communities is through government restrictions, taxes, fees, or bans. These drive large-scale change faster than trying to persuade people to change their habits, or lengthy campaigns against corporations that are financially wed to expanding plastic. Cities that ban or tax plastic bags see an immediate environmental benefit and savings for the municipality, which no longer has to dispose of these notoriously unrecyclable products. Surfrider Foundation and Greenpeace have online tool kits for working to ban plastic bags in your community. Start at the local level, with your city council, and then, once you win there, aim for statewide restrictions on plastic, which are easier to achieve if there is a groundswell of support in local communities. While you're working on city and state bans, you can also get bans on bottled water, plastic bags, and other unnecessary plastic at your school, workplace, and church. This is easier and a great way to hone your arguments and organizing skills!

And, because 99 percent of plastics are produced from chemicals made from fossil fuels, the ultimate way to reduce plastic is to reduce the availability of fossil fuels. Right now, the fossil fuel and petrochemical corporations are planning a massive build-out of hundreds of new plastic production facilities, fed by the fracking boom in the United States. We have more fossil fuel projects in operation than the climate can sustain, and we already have more plastic pollution than we can manage. Working on campaigns to stop new fossil fuel extraction helps slow plastics production.

"We need systemic change here, and it starts by holding plastic and fossil fuel corporations accountable for the impacts of their harmful products and practices. We want to make sure we're voting for the people who are really going to make that happen. It's good to make lifestyle changes that reduce our carbon footprint because it's very important to practice our values. But we also have to take the next step and join the political fight because the fossil fuel industry is not going down without a fight. We're in the middle of it."

Iain Armitage speaks.

CHAPTER TEN

Food, Agriculture, and Climate Change

It was November 28, Thanksgiving, and it seemed fitting that our focus that week was food and agriculture. We usually get together for Thanksgiving, often at my home, but this year my "home" was an extended-stay hotel room in D.C. Our usual eclectic clan made up of my children (adopted, biological, and step), grandkids, friends, and ex-husbands and their wives and exes and their children were largely missing. Three of the four kids, however, did come: Mary "Lulu" Williams, who spent all four months in D.C. with me, helping with Fire Drill Fridays, and Nathalie Vadim, the firstborn of my first husband who has been part of my family for many decades, joined us from New Mexico. My daughter, Vanessa Vadim, who had come down twice already, engaged in civil disobedience, and gotten arrested, was here again with my grandchildren, seventeen-year-old Viva and twenty-year-old Malcolm. Both of them wanted to join Fire Drill Fridays.

It being a holiday, the Greenpeace office was unavailable for the teach-in, but Ira Arlook and Karen Nussbaum invited my family and our documentary film crew to join their family for Thanksgiving dinner at their house, and we set up the teach-in in their living room.

Looking back over my time in D.C., I realize how important it was for me to have had Ira and Karen close by. Ira, as the person in charge of press for Fire Drill Fridays, was the one I spent the most time with, my companion on the train rides to and from New York, and at my side every day as we did back-to-back interviews during the first two months. Karen was the comforting face I could always find in the crowd on Fridays, who did civil disobedience three times and spent a night in jail and who helped us strategize on ways to bring the labor and climate movements closer together. I had known them both since 1972, and we had worked together in a number of successful movements—the anti–Vietnam War; 9to5: National Association of Working Women, which had inspired my film **9 to 5**; and most recently Working America, the labor movement's field organizing arm, which I view as one of the best ways to get a progressive Democrat elected. As a newcomer to the world of D.C. movements and an older woman from a totally different milieu working with a team of unknown and differently seasoned young people, having familiar comrades there to share time with, reminisce with, go out to dinner and the movies with, was grounding.

It was generous of them to open their home to the Thanksgiving commotion of having eleven extra people to cook for, especially given that Karen insisted on doing all the cooking. She's gifted that way.

Our other guests that evening were Jim Goodman, who, with his wife, Rebecca, ran a forty-five-cow organic dairy and direct-market beef farm in southwest Wisconsin for forty years, and Ricardo Salvador, a senior scientist and director of the Food and Environment Program at the Union of Concerned Scientists.

The Teach-In

I was grateful and amazed that Jim and Rebecca had come all the way from Wisconsin on a holiday to be part of Fire Drill Friday. Jim grew up on a farm that his family had purchased in 1848 and credits more than 150 years of failed farm and social policy as his motivation to advocate for a farmer-coordinated and farmer-controlled consumer-oriented food policy.

At the teach-in, I asked Jim what it was like growing up in rural America and how it's changed.

"When I was a kid, you had a lot of small towns with small businesses, machine dealers, car dealers, grocery stores, movie theaters. It used to work. That's gone now because the tax base is gone, and the people are gone. There are no kids anymore. The school districts are closing down. So we've basically decided that rural America just doesn't need to exist anymore. I grew up seeing small farms, probably a dozen dairy farms between our home and town. Now there's one left. The farms are still there, but they are feedlot farms or they're selling to farms that are milking thousands of cows. We somehow have a mandate to feed the world. I'm not sure who told us to do that, because I think the world is very capable of feeding themselves if we let them. Small farms, especially if you look at the developing world, are what produce most of the food in the world. We have to go back to the regenerative agriculture system that these small peasant farms have been doing for centuries."

"I've read that farm bankruptcies are up 24 percent. Is this true?" I asked Jim.

"Yes, farmers are in crisis. We're seeing incredible crop losses due to more heat, more floods, more droughts, and more pest invasions. You see, since the 1980s, farms have been getting bigger, and they're growing mono-crops of genetically modified corn and soybean, food for livestock, not food for people, and that's really hard on the environment.

"The kind of farming I did, organic farming, regenerative agriculture that peasants all over the world do, you don't have to have a lot of inputs. You manage your crops and your livestock so it's a holistic cycle that continues to feed itself. True, you don't get the giant yields, but the value of the food, the quality, the nutritional value, is much better, it's produced locally, and it's culturally appropriate for people. The only thing it doesn't do is make a lot of money for corporate agriculture.

"With the genetically modified mono-crops of corn and soybeans, on the other hand, you continually rotate those two crops, which means that the soil holds the roots, but you have to provide everything else: the fossil-fuel-dependent fertilizer, the water, the pesticides. Industrial agriculture systems appear to be productive, but they are actually biological deserts where the definition of a desert has to do with lack of life and lack of diversity, lack of biodiversity. Pests have traditionally been controlled by extreme cold during the winter seasons. These days, because of the fact that we don't have extreme cold sufficient to actually kill off insects in the winter, their populations now are propagating to greater and greater numbers year after year. What that does in the industrial system is ratchet up the need to get more inputs either through the utilization of insect-resistant crops or through the use of more insecticides, more pesticides. And pesticides in particular tend to be the more noxious of the chemicals that we use in agriculture.

"Climate change is making it more difficult to produce, it is making it more expensive to produce, it is making it riskier, and that means the cost to the public is increasing. There's a body of evidence showing that the nutritional value of crops is diminishing under climate change. One model that is likely to explain it, or at least is going to be involved, is that with high temperatures you get faster crop growth, and it gives crops less time to actually absorb nutrients, and that's called the dilution effect."

Ricardo jumped in.

"Rather than relying on the natural cycle of what we call regenerative agriculture, where soil holds the water and nutrients that your crop needs, industrial agriculture, requiring massive outputs for profit, needs massive inputs of fertilizer, water, pesticides. This kind of factory farming is productive on one level. It produces a lot, and that's how our current economic system judges it: by quantity of output.

"Back when this nation was colonized and there were about a billion people on earth, we saw the planet as an infinite source of resources, an infinite sink for all of our wastes. And we developed our economic thinking with those assumptions. Now there are almost eight billion of us, and we're discovering that the planet is finite in terms of the resources, that it is not an infinite sink for our waste. Yet our economic thinking hasn't changed. This means you can degrade the soil. You can extract resources from the ground that will never be replaced. You can pollute the water. You can pollute the air. But if you produce a lot in terms of wheat, cotton, rice, corn, soy, we say that we're being successful. All of those things—the pollution, the degradation—are not being measured. We're operating with very powerful twenty-first-century technology, but we haven't updated our economic eighteenth-century thinking. And climate change exacerbates all of this.

"Yet farmers can be part of the solution. They could be capturing carbon rather than releasing it out into the atmosphere. They could actually put as much carbon as possible into the deep roots systems, reincorporating organic matter as much as possible, making sure that we're closing nutrient cycles whereby livestock and crops are working together in order to be able to do that. Livestock eat the crop. They generate manure, which farmers then reincorporate. And there is an actual science as well as an art in terms of knowing how to do that. Right now, we've broken apart all of those systems and created linear systems where the nutrients are not regenerated. They're simply purchased from someplace else. The phosphate comes mostly from northwestern Africa, for instance. The potassium may come from southern Canada or from Florida.

"So that means that when we run into problems because of the fact that these soils are now impoverished because there's very little organic matter given, it will wash away quickly. They flood easily. You can't get machinery into them. You can't plant. That means that you are delayed. If you don't get to plant at all, then that means that you have no livelihood for that particular year unless you're on government programs. So it's essentially a disconnect that comes from pretending that we're running a factory that just happens to be outdoors except that unlike a factory we don't control most of what's happening outdoors. So we've not valued the knowledge that we need in order to be able to work with natural mechanisms. We have an agriculture model which turns out to be very destructive, inimical to humanity's long-term interest. But we have alternatives. There are better ways of doing this. And we're not investing in those ways. And we're not telling farmers about them, because there is no profit in that. So the private sector, the corporate sector, is not motivated to get into that type of agriculture."

"Ricardo," I said, "we've become so accustomed to thinking of carbon as bad, but now you're saying we want carbon to go into the soil. Explain."

"These days when we talk about carbon concentration being high in the atmosphere, this means that we're getting elevated levels of carbon somewhere that it did not used to be. But the carbon itself is part of a natural cycle. We are carbon-based organisms. The carbon cycles through the entire mantle of the planet, throughout the oceans. And the issue is that because of the way we burn fossil fuels that used to be stored in the ground, we are releasing tremendous amounts of carbon into the atmosphere that did not used to be there. That's referred to as anthropogenic carbon." **Ah yes, carbon that comes from a cemetery of prehistoric sea creatures that we rename fossil fuels, as the biologist and poet Sandra Steingraber said back on the very first Fire Drill Friday.** "That's actually the problem. And this reliance on fossil-fuel-based inputs is justified by saying that because of poverty on the planet we need to produce a lot more to feed the world. In fact, we need to be producing a lot less. And we need to produce the right stuff. If we returned to regenerative agriculture, it would be better for the planet in terms of the carbon cycle, in terms of our own well-being. We're talking about a phenomenon here—climate change—that humanity has never encountered before. Therefore, most of us are not prepared to even believe it. Because if what science is telling us about what climate change is doing to us, in the very short term, is true, that means that our current lifestyle will no longer be viable. And that's too much for us to accept. It's much easier to deny that."

I tried to find the faces of my grandkids to see how they were taking this news, but they were listening from the kitchen. I was happy to have my daughter, Vanessa, who has twenty years of experience in rural and urban food production and is currently farming twenty-seven acres in southern Vermont, as part of the teach-in. "Vanessa, do you plant your perennials to help capture carbon in the soil? Is that what has led you to focus so much on perennials on your farm?"

"My growing is perennial oriented, but it's also systems oriented. With annuals, you have to replant every year, which means disturbing the topsoil, roots never getting the chance to develop enough to tap down and hold soil in place or access deeper water or outcompete weeds. It means more inputs—water, fertilizer, herbicides, more topsoil loss, less resilience.

The original prairie grasses of the Midwest were perennial with very deep root systems. Great herds of
bison, the indigenous ungulates of North America,
roamed wild and aerated the soil with their hooves,
which helped keep the soil and grasses healthy. They
literally roamed, whereas cows will stay in one spot
and eat the grasses down to the roots. Along with the
shift in grazing, we started large-scale tilling and replacing perennials with annuals. These were the main
causes of the dust bowl that drove thousands of farmers and their families to abandon their land. We think
of deforestation as causing desertification, but it is
just as much loss of grasses and other perennials that
is at issue. It was the environmental devastation of
the dust bowl that was the catalyst for the New Deal.
Let's make our current food-systems crises work for us
in igniting a Green New Deal.

"The abandonment of perennials is in part the
change in patent law. It used to be that you couldn't
'own' a living organism. But now that corporations
are allowed to own seed, it is profitable for Monsanto
et al. to sell their seed every year. And not just annuals,
but annuals that can't be reproduced—leaving behind
the centuries-old tradition of seed saving. Farmers are
being sued if they try to replant rather than buying
new seeds each year. Today, at least 75 percent or more
of our calories now comes from annual crops, even
though annuals are input intensive and can cause soil
degradation when mono-cropped and heavily tilled.

"We have solutions, but they require us to go back

to a relationship with the land and food that reflects native ecosystems and means much larger numbers of people actively engaged in the farming process."

I felt proud listening to all that my daughter knows about these things. I wished my father, a product of the Great Plains of Nebraska, were still alive to hear her. A farmer at heart, he was always happiest with his hands in the dirt. To just about his dying day, he kept beehives, chickens, and compost bins at his home in Bel Air, where it was unlawful, but his sympathetic neighbors remained mum despite the early-morning rooster. You can take the man out of Nebraska, but you can't ever take Nebraska out of the man.

Jim added, "But, Jane, farmers can be part of the solution to climate change. We can transform how we farm. We can reduce emissions, keep carbon in the ground and in the soil, not in the atmosphere, and that would be better for the climate, better for farm-workers, better for the planet, better for the health of all of us."

"I've read that three-fifths of agricultural land is used for cattle and that cattle produce methane," I said to Jim. "Should we stop eating red meat, or is there a way to farm cattle that will not be harmful? Or is it just the amount of meat that we eat?"

"It's the way we raise meat, yes. And we don't need to eat meat three times a day; we don't need to eat meat three times a week," Jim replied. "I think every-body should eat less meat, but that's a personal choice. Small farms may have a pig or a cow, a few chickens, and that's a really important part of that farm because it provides a protein. It provides fertilizer for the soil, and it's very low impact. But we raise beef cattle in this country in confined operations, feed them nothing but corn and soybeans somewhere in the middle of Kansas, Texas, Nebraska, and then they're shipped somewhere else to be slaughtered, which means there's a lot of transportation involved, there's a lot of monoculture crops and clear-cutting involved, and ruminants like cattle produce the heat-trapping greenhouse gas, methane emissions, through digestion, and all this produces a kind of meat that's not good for people to eat because it's not grass fed. We need to stop that type of production and go back to a much smaller scale of livestock that's integrated into the farming system."

Hearing Jim made me remember driving into the plains in Colorado during filming of **Our Souls at Night** with Robert Redford. You could smell the feedlots long before you saw them. And once you saw them, it was the best possible motivation for giving up meat: cattle standing knee-deep in their own excrement as far as you could see, packed together so tight they were barely able to move. Agrobusiness isn't just bad for people; it represents cruelty to animals. Then I remembered a totally opposite vision, what you see when you take a train through Europe or drive through Argentina or even Vermont: cattle grazing on grass on small farms.

REGENERATIVE AGRICULTURE

Ricardo Salvador

"Regenerative agriculture means we pattern what happens in the natural world. Nature built up very thick layers of very fertile soil in relatively short periods of time—geologically speaking—tens of thousands of years. And the way that that happened by natural cycles was that vegetation would grow during warm parts of the year and then would die back during the cool parts of the year. That vegetation would be broken down by fungi and bacteria, would be reincorporated into the soil, and in the soil that vegetation becomes what we call organic matter. The organic matter serves a lot of very important functions. But just at the physical level, you can think of it as being glue between fine particles of soil that otherwise would disperse. They would not clump together. And the fact that soil clumps together in particular ways—what scientists call structure—means that soil can do a magical thing. It can simultaneously do two things that appear to be the opposite of each other. First of all, it can hold plant-available water so that crops can take it up and transpire. That's how they keep themselves cool and they take up nutrients in that stream of water. But the other thing that soils can do is drain water so that plants are not standing in water because roots need to breathe. If they were flooded, they're not able to breathe, and so the plant doesn't function. It dies off in a matter of just a few days. The fact that soils can do that is due to organic matter, which actually functions as a glue. But more than that, it increases the fertility of the soil because the properties at the chemical level are such that minerals that are constantly breaking down are actually held, they're bound chemically to that organic matter, and the entire system regenerates itself. It builds itself. Over the 150-plus years since we've been farming in the industrial mode, we've not only interrupted that cycle so that those fertile soils are no longer being created; we've actually degraded them. In many places—and the U.S. Midwest is very good evidence—we've actually reduced the depth of the soil by 50 percent.

"So, what nature did over 10 to 14,000 years, we've undone in a period of about 150 years. And we mask that. Most people aren't aware of it and actually argue against the type of system that Jim and I are describing because they say, 'All you need to do is to buy the fertilizer and nutrients. You don't need the soil to be doing that work for you.' And so that discounts the value of what nature does and makes farming basically a transactional sort of industrial operation. And as I mentioned earlier, because of the fact that this kind of a system values productivity measured by capital investment and return on that investment, then that appears to be efficient, but it's not measuring the degradation of the soil, the loss of biodiversity, the effects on climate, people's declining health.

"What we are producing is essentially stuff that goes into processed food that makes us sick, it pollutes the environment, pollutes the water, generates greenhouse gases. But as I've mentioned earlier, it doesn't have to be this way. We have better models that not only have been practiced in the past, but which nature actually models for us. That's what regenerative agriculture is about."

"Right," Jim added. "As an organic dairy farmer, part of the requirement is that the cattle be pastured. It only makes sense to pasture cattle. Cattle like to be in pasture; that's what they're supposed to do. And now they just warehouse them with these pumps."

When the teach-in was over, I thanked Ira, Karen, and my guest experts and headed home to try to rest up for the next day's rally. It would be Black Friday.

The Rally

The first speaker at the rally was Kallan Benson, a really smart, homeschooled fifteen-year-old member of the Sunrise Movement whom I had gotten to know. Kallan was dressed in gauzy black funeral clothes.

"I'm a long-term volunteer at the Smithsonian Environmental Research Center. I work alongside a study that has been looking at the impact of climate change on the nutrition in our food. The way our food gets nutrition is that the plants take up water from their roots, which brings up nutrients with it. On a

Kallan Benson, "in mourning," speaks at the rally.

plant's leaves there's something called a stomata cell. That cell does the respiration for the plants. It brings in CO_2 and it lets out water. In an increased CO_2 environment, it opens up less often. That means it lets out less water, which means, internally, it doesn't need to draw up as much water. That means that our food becomes less nutritious, we lose nutrients in our food. The other thing is that it has a drastic impact on insects, which creates an even bigger problem because insects pollinate our food. This has to end. We students have over three hundred strikes across the country every week. But we need you because you can also do more. As youth, we cannot engage in our political system because we can't vote, so we need you to vote and to go to your local officials, your state-level officials, your federal officials. Today I wear black because I am in mourning. I am mourning our future, and I am mourning what we have already lost. We are having a funeral for the future because today is Black Friday, a day of consumerism."

Sarah Schumann spoke next. She is a commercial fisherman who splits her season between the waters of Rhode Island and those of Alaska. "Climate change is increasing the risk and the uncertainty associated with being a commercial fisherman. Today in Maine, warming waters have canceled the last six shrimp seasons. In California, warming waters have led to toxic algae blooms that have cut short the crab season and led anguished commercial fishermen to sue fossil fuel companies over lost income. And in Alaska, a marine

Sarah Schumann speaks.

heat wave has depressed some cod and salmon populations, leading to a fisheries emergency.

"Industrial offshore wind farms, some of them owned by oil companies, are racing through the permitting process at breakneck speed. Within a decade, hundreds of towering turbines will stretch across our fishing grounds, and hundreds of miles of electrical cable will crisscross seafloor habitats. Once installed, these turbines will supply millions of homes and businesses with emissions-free energy. But what kind of ecological consequences will this unprecedented ocean industrialization have on the marine wildlife that sustains fishing economies? People like us should be leaders on climate solutions, not casualties. Humanity does not have a choice about whether we address the climate crisis, but we do have a choice about how we do it. Do we want energy policies that are directed by people in suits, or should they also be driven by people in boots? Do we want a climate action agenda rooted in fear that leads us to grasp at any apparent techno fix no matter how destructive it may be in its own right? Or do we want a climate action agenda rooted in faith? Faith that human beings can rise to the situation and can tackle this monumental challenge. This moment in time isn't just about reducing the carbon molecules in the atmosphere. It's about redefining ourselves by the things that truly matter. The things that feed our bodies, the things that feed our souls, and the things that feed our future. Harvesting wild seafood on family-owned fishing boats and raising crops on family-owned farms does all three. These things cannot be collateral damage of the climate crisis or of the choices that we make to address this crisis."

I could see some climate activists in the crowd wincing as Sarah critiqued industrial wind farms, but by then I'd done enough reading about what has worked in various parts of the country when all parties involved are at the table together developing a way forward. Everyone may not get 100 percent of what they want, but everyone has had a say in the process and understands how and why decisions were reached. That's what the Green New Deal calls for. Within that framework, oil-company-owned wind turbines wouldn't ride roughshod over the interests of local fishing industries.

A VOICE FROM
FLY-OVER COUNTRY

Jim Goodman

"It's time to invest in regenerative agriculture and end our investment in the agricultural industrial complex, one of consolidation, factory farms, and junk food. It's time for agriculture that protects and restores the biodiversity of our land and the oceans; an agriculture that promotes organic farming based on healthy soil, healthy food, and healthy people; an agriculture that respects the rights of farmers, farmworkers, and nature, a food system controlled by farmers and the people we feed, not corporate profit margins or the global economy; an agriculture that respects the rights, the treaties, the lands, the culture, and the knowledge of indigenous peoples all over the world, here in the U.S., in Brazil, Bolivia, Africa. People who protected for generations their seeds, their knowledge, and passed it on to us.

"It's time for a Green New Deal. The Green New Deal is not, as some would call

Jim Goodman speaks.

it, the 'green dream' or whatever. We know exactly what the hell it is. And we needed it yesterday. Only a plan like this, a plan for overall societal reform, can work. And the genius of the Green New Deal is that because it calls on grassroots action to move beyond the greed of corporate America, it brings every one of us, at least every one of us in the 99 percent, together. It calls on every-one who wants fair wages and a fair econ-omy, everyone who wants equal education, everyone who wants racial justice, everyone who wants gender equity, everyone who wants real health care, everyone who wants clean air, clean water. And perhaps especially farmers like me from fly-over country who only ask for fair prices, fair markets, and a chance to show that it's not time to get big or get out.

"Resistance is an act of necessity. We can only fight the inaction of government, the inaction of those who don't believe in science, the inaction of all those who are sleepwalking toward extinction, by putting ourselves on the line. It's time. Some of you may remember the Free Speech Movement in the mid-1960s. I'm paraphrasing Mario Savio who said that there comes a time when the operation and the machine becomes so odious that you have to throw yourself in the gears and make it stop. It's time. It's time to make it stop and start over for all of us."

When the rally was over, people moved to the Capitol steps for the civil disobedience while I watched from a distance with Vanessa and eleven-year-old Iain Armitage, star of **Young Sheldon**, on holiday break, who had come from Delaware with his mother to speak at the rally. It was a big deal for me to watch my grandchildren standing there on the steps, chanting,

What do we want?
Climate justice!
When do we want it?
Now!

I had brought Malcolm with me to Toronto a few years earlier for a big Jobs & Climate march and rally, and both kids had been with me at a rally against the North Dakota pipeline, but they seemed more engaged now, more aware that it was their future they were protesting on behalf of. I managed to get hold of a megaphone so that when each of them was handcuffed and led away by the police, I could be the proud grandmother and shout their names, cheering them on.

Eventually, Vanessa showed up with Viva, freshly released from juvenile detention. She was tired and radiant, proud that she'd had the guts to go through with it despite the team's admonitions that she was too young. She commented on how dehumanizing the prison system seemed and on how aware she was of her privilege.

"My cellmate was seventeen years told," she told me. "In for armed grand theft auto, who had been in juvie twice already, and told me she didn't expect to see someone like me in there. She didn't seem to care when I told her I was arrested for civil disobedience over the climate emergency. It was definitely eye-opening, empowering, and uncomfortable. Even though I understand about civil disobedience, I felt a bit ignorant for going to jail on purpose when there were so many young girls around me who were trying everything to get out."

Vanessa, Viva, and I were all there, emotional and jubilant, to receive Malcolm when he was finally released. Malcolm, unlike Viva, is reserved and quiet, but when I saw his beaming face, I knew that it had been a positive and memorable experience for him.

Jane stands with her daughter, Vanessa Vadim, and grandchildren Malcolm Vadim and Viva Vadim outside jail, just after their release.

What Can I Do?

Food is one of the easiest and most direct ways we can do something about the climate crisis, starting with what you eat and extending out to the food and agricultural policies you support.

Reducing the amount of meat and dairy you consume is a win-win for your health and the health of the planet. Eating more fruits, vegetables, and whole grains reduces heart disease, diabetes, stroke, and certain types of cancers. Studies have shown an organic diet quickly reduces pesticide residues in our bodies. Overall, eating a mainly plant-based diet of organic and petrochemical-free foods is healthier for you, for farmworkers, for other mammals and birds and insects, for the soil, for the water, for the air—for the earth!

On top of individual consumption, we can also work together to demand institutions like company and school cafeterias and grocery stores shift away from industrial-scale agriculture that emits tons of greenhouse gases. From industrial meat farms producing lakes of animal waste, to forests cut down to grow soy feed for animals, to the synthetic fertilizers dumped on crops, to the trucks that transport global commodities (like soy and corn), to the airplanes that fly salmon caught off Alaska to China for processing and packaging, then back to the United States, the more we can eat locally, less chemical-dependent, less processed, less packaged food, the better for us and for the planet. Besides changing how you shop, you can ask your retailers and restaurants to reduce their purchases of industrial meat and dairy and to provide more plant-based options, source ethically and ecologically produced meat, and buy from local growers and suppliers.

Reducing food waste is another important step for the climate. So much food is wasted. We all know the experience of tossing that now-rotten bag of spinach in the trash after two weeks because we just never managed to cook it in time. Food waste is not just a waste of valuable nutrients; when it rots in a landfill, it releases methane emissions, another driver of climate change.

No judgment if you never got to that spinach; just make sure you compost it instead of throwing it in the garbage, so the organic matter goes back into the soil. It's easy to compost at home in the backyard or, if you live in an apartment, try vermicomposting—composting with worms. (I know that may sound weird, but it is amazing how well it works and how those little creatures can make food waste disappear.) There are lots of cities and states—from California to New York—that run citywide composting programs. When people have curbside compost pickup, just like trash and recyclables, there really is no reason not to compost.

Food can become part of your politics. Insist that your city, state, and federal politicians stop subsidizing factory farms and instead support small organic farmers. Write to your elected officials, and ask candidates running for office to end subsidies for industrial meat and dairy products. Advocate that they promote the production of healthy fruits and vegetables from local farms and cleaner, more sustainably produced meat and dairy from ecological livestock producers.

There are many fantastic organizations you can join that are already working hard to ensure a climate-conscious food system. Food & Water Watch runs a campaign to end factory farming. Greenpeace's Less Is More campaign and Friends of the Earth also have climate-friendly food campaigns.

If you can lend support to grassroots and worker-led organizations that fight for economic and health justice, please do that too! Some I love are United Farm Workers, the only union of farmworkers in the entire country, the Coalition of Immokalee Workers' Fair Food Program, Migrant Justice's Milk with Dignity campaign, and Restaurant Opportunities Centers United's fight for a fair wage. Milk with Dignity and the Fair Food Program pressure corporate growers and retailers who pay poverty-level wages and commit environmental destruction.

Directing actions and campaigns toward corporations is genius for at least two reasons. One, it shifts the burden from individual consumers, who can only do so much on their own, onto wealthy corporations that can certainly afford to pay farm and food service workers more. Two, corporations are much quicker to respond to public and consumer pressure when it comes from a groundswell of support rather than from scattered efforts here and there. On an international scale, you can also become a member of groups like Food First and La Via Campesina.

The march.

CHAPTER ELEVEN

Climate, Migration, and Human Rights

By the first week in December, Annie and I both knew it was time to start talking about growing Fire Drill Fridays beyond D.C. Clearly this action had touched a need people were feeling: the desire to put their bodies on the line for the climate. Folks from around the country and around the world were asking how to bring Fire Drill Fridays to their towns and cities. We needed to try to make it happen.

So when Annie and Maddy asked me if I wanted Fire Drill Fridays to become a Greenpeace project, I answered with an unequivocal "yes!" Historically, Greenpeace is associated with brave, bold, large-scale actions that draw international attention to critical environment issues—blocking Royal Dutch Shell's big oil rig from leaving Seattle's harbor to drill in the Arctic, forcing Burberry to stop using toxic chemicals in its manufacturing, exposing nuclear testing near Alaska—often actions that, by their very nature, appeal to frontline, risk-taking types while millions of others showed their support through donations.

Fire Drill Fridays offered Greenpeace a campaign that could build a grassroots movement of people new to civil disobedience, the exact demographic we had aimed to inspire from the outset. Both Annie and I sensed that civil disobedience needed to become the "new norm" as the climate crisis loomed, and Greenpeace, with a few additional hires, had the know-how and capacity to scale up this effort.

We convened a meeting in the Greenpeace office again—in the same room that served as our weekly teach-in set but now in its usual form with a big meeting table to gather around. Annie and I were joined by Janet Redman, who leads Greenpeace's climate campaign, and Jose Martinez-Diaz, who leads Greenpeace's work to engage millions of Greenpeace members around the country in campaigns for change. We spent the whole afternoon strategizing about how to continue to build Fire Drill Fridays, leading up to the next election, when it is so crucial we elect a climate leader, and then continuing afterward to ensure that the president—whoever it is—acts with the level of boldness that both the science and a growing percentage of the public demand.

It was exciting to think about this idea, hatched a mere three months earlier, going national.

This first Fire Drill in December was our ninth,

and it focused on migration and human rights. In addition, Climate Action Day was being celebrated, when people all around the world demonstrate to demand that their governments take climate change seriously. As a result, robust numbers of participants were coming to Washington, D.C., and the speakers signing up for our Friday rally were impressive, and many more than we could handle. We also had several organizations that wanted to partner with us for the march that would join our rally, so many that we decided to move the Fire Drill from the Capitol to Franklin Square. The square was an appropriate marching distance to the D.C. office of BlackRock, the world's largest asset-management firm, where we had planned our civil disobedience.

The Teach-In

At that Thursday's teach-in, my guests were the Reverend Fletcher Harper, an Episcopalian priest and executive director of GreenFaith, a global multi-faith climate organization, who has been a leading voice in the faith community's response to the climate crisis for the last two decades, and Saket Soni, a labor organizer and human rights strategist.

Saket is the founder and director of Resilience Force, a national initiative to transform America's response to national disasters by making America's workforce stronger and more resilient. Together these

Saket Soni speaks.

two thinkers and activists were working daily with the refugee population dislodged from their homes by climate change, and I knew they would have insights that would illuminate this issue for those who were new to the idea of climate refugees.

All over the world as water dries up, farmland turns to desert, coastal areas flood, and permafrost melts, ecosystems can no longer support the communities they once did. And experts say it's going to get much worse in the coming years. This is both a climate crisis and a human rights crisis. Families fleeing land that has become uninhabitable do not get warm welcomes elsewhere. Migrants routinely face life-threatening hardships, discrimination, and repression as they search for safety for their families.

As far back as 1990, the Intergovernmental Panel on Climate Change noted that the greatest single impact of climate change could be human migration. And we're seeing this projection come true. The latest World Health Organization estimate predicts that as many as 200 million climate refugees will be on the move by 2050! The UN Refugee Agency reports that right now 71 million people have been forced from home, more than half of whom are under the age of eighteen. This is the highest number of displaced persons the agency has ever recorded. Along with apocalyptic fires, floods, and rising sea levels, the increasing numbers of climate refugees have been forcing more and more people to acknowledge that climate crisis is not a future danger but a here-and-now reality.

As these numbers increase, societies have choices to make about how they address this human catastrophe. One of the tragedies of this moment is that the climate crisis is a collective crisis that requires a collective solution at a time when the notion of the collective, the common good, the public sphere, is being eroded. I realized that part of the fight against the climate crisis means rebuilding community.

The climate emergency tests who we are as a country, what we value, and what we support. So many people were shocked and saddened by that photo of an immigrant child who drowned in the Mediterranean as his family fled Syria's drought and war, and another image in 2019 of a father and daughter facedown at the banks of the Rio Grande. These images capture the consequences of the dangerous journeys families undertake when their lives at home become unsustainable. Do we decide to erect taller walls and enact stronger immigration policies to keep people out and protect what we see as ours? Or do we open our arms to embrace the strength of immigrants who come seeking to build new lives here, as the majority of our families have done in decades before?

We've got to do better as a global society.

During two Republican administrations—Ronald Reagan's and George H. W. Bush's—the approach to immigration and refugees was very different. They abolished immigration prisons altogether, and pilot projects that offered support to migrants consistently showed success.

I was struck by how many times during the teach-in, as well as during the rally, Reverend Fletcher and many others spoke of the need for cultural change, and, I have to admit, at the time they said this, it made me antsy. Cultural change is slow. The crisis requires new policies and laws—fast. By the end of the day, however, and after hearing the immigrants and religious speakers at the rally, I realized that while we urgently need a new, more humane immigration policy, we simultaneously need to change our feelings about immigrants. That's cultural. If we gain good policies but hatred and resentment continue to fester in too many Americans, we remain just one bullying, dog-whistling autocratic president away from climate apartheid.

At the teach-in, Reverend Fletcher talked about the climate refugees from a biblical perspective. He noted that the Bible is full of stories about refugees: Moses, Abraham, and Jesus.

"It's stunning to me how many times the Bible talks about protecting the well-being, the rights, and dignity of migrants. And so now the climate crisis is accelerating and driving the numbers of migrants, climate refugees, higher and higher and higher. And we see, on the one hand, the world's wealthiest and most powerful people who are choosing to pay for private firefighters to protect their homes, while regular people have to uproot and move.

"We're seeing the wealthiest and most powerful investment houses in the world continue to plow money into fossil fuel infrastructure, and then when people are displaced and face punitive, legal, and military repression in places where they try to move, these same firms are investing in private prisons and detention systems. They're profiting by causing the problem, and they're profiting at the other end by mistreating the people who are forced, through no will of their own, to have to move.

"There is a fundamental moral-spiritual decision that we have to make as a human family: Are we going to welcome people who are displaced? Are we going to love them? Are we going to create communities that celebrate the gifts that they have to offer? Are we going to ensure that they have a right to education and to good work and to safe living conditions? And are we going to do that together as a human family?"

I asked my other guest, Saket, to describe his experiences working directly with refugees, including some who had been displaced in the United States by big climate disasters like Hurricane Katrina.

"I guess I would start by saying that the phrase 'climate refugees,' when you hear it, probably evokes someone far away," Saket said. "But this issue is not a faraway problem. Most of my time is spent with people displaced from climate change here in the U.S. Some of them might not call themselves refugees. All of them would consider that they've been turned into strangers in their own land. They've lost their homes. Many have moved away from the community they lived in, called home. Many will never come back. But they were born and raised in the U.S. They felt like things were stable. Things were going well for them. They didn't ever imagine moving away. What if that climate refugee was you? What if it wasn't someone far away? What if it wasn't someone who was crossing an international border?"

Saket described his work after Hurricane Katrina as an organizer in New Orleans's City Park. This splendid garden, "one of the most beautiful places on earth," he said, became a makeshift labor camp after Katrina.

"And there was a woman there who had been hit by a hurricane in Arkansas. She lived in a FEMA trailer. And when Hurricane Katrina hit New Orleans, FEMA came to her trailer in Arkansas and said, 'You got to get out. We need these trailers.' She followed her trailer to New Orleans because she had nowhere else to live. She was born and raised in the United States of America. And she was working as a plumber after Hurricane Katrina, living in a tent in City Park."

In the last decade as hurricanes, floods, and fires have become more frequent and more destructive, large portions of cities, entire regions, have been destroyed. "Most of the rebuilding is done by un-documented immigrant workers who've come to the United States," Saket said. "Many of these workers are rebuilding the homes of people who previously might have voted for their deportation. The greatest paradox of all of this is that in the Florida Panhandle—in fact, in Bay County, the very place that President Trump went and asked at a rally what do you do with the immigrants and received a response, 'Shoot them!' In that very place, where seven out of ten people voted with the president on immigration, in that very place, everyone, without exception, is getting their homes rebuilt by an undocumented immigrant."

Saket saw this as an opportunity for immigrants, though. "We're getting in there and taking that op-portunity to build bridges. If you help me in the hardest period of my life, then you become part of my thought process as I'm putting together my new priorities for the rest of my life. And I'm going to pri-oritize loving you if someone comes in and helps me make a bond. So we're trying to build bonds between immigrants and nonimmigrants in the place that it's hardest and where it hurts the most so that people change their minds about immigrants after going through this."

"And this is really happening?" I asked, astonished.

"It's happening. We rebuilt the home of an elderly couple whose son told us he put up a sign that said, 'Looters Will Be Shot.' The ultimate protection when nothing else is there to protect you is the sign you scroll with your own Sharpie and you put in front of your door. Even if you don't have a gun, you declare, 'Looters will be shot.' Well, we rebuilt that home. And after we had dinner with that family, they threw away that sign. So there is a lot of hope there. But I think the way that we're doing that is not by sitting down and debating with them the finer points of immigration policy or climate change. It's by helping. It's by being there. It's by building bonds. And it's by working on a problem that's more important than hating each other. It's by working on a problem that we can all fix. The solution is both attacking the root causes of climate change and also living in a more adaptive world, living in a world where we're not strangers to each other, living in a world we rebuild together.

"It may sound odd to say," Saket continued, "but in many ways, I think this is the heart of the matter. If we don't have these bonds, then we do not have the ability to create change. I mean, certainly, elections matter enormously. But the long trajectory is that when you build these kinds of values-based relationships at the local and regional and national level, you can change the world. It takes time, but you start by getting to know your neighbors."

Reverend Fletcher also talked about the rebirth of community as necessary, and as possible, because of this crisis. Exactly the sentiments I had been feeling.

"I feel strongly that the resolution that's set forth in the Green New Deal is a tremendously important foundation for what needs to happen. Because what it says is that we need a response that is based on the values of commitment to those communities that are on the front lines, a substantial investment in their leadership and in their well-being and in their resilience, a commitment to treating the workers who are going to be doing all the work of building this new country and this new world that we need, that their rights need to be protected, and that there's an urgency and that this needs to be done at scale and that the government needs to play a big role to help prime the pump for the rest of the system to get going. I think the fact that the Green New Deal has begun to go to work on the public's imagination, that's a very important starting place."

"Yes," I responded. "And it's also what builds resilience in a community. And God knows we're going to need resilience."

Saket added, "I think that it's not just about welcoming people to America. It's about realizing that we actually need a new America. We need to rebuild America. We need a bottom-up rebuilding of America that's better, that's more resilient, that's more loving. And that's what we're all going to do together, the people like us who are historically migrants and the new ones."

Reverend Fletcher added, "We're not ahead of the curve on this, and so we have to be doing the proactive work to prevent further problems in the first place. On a very broad level, faith leaders have to talk about a world that makes it a lot easier for people who are forced to move to do so safely, humanely, with dignity, with proper levels of support. We need to be doing the work culturally to prepare for that because that's coming."

When we started taking questions from our listeners, a woman in California who had lost her home in a fire asked what could be done to prevent such things in the future.

"Build a movement," Saket said without hesitation. "A movement to make PG&E a public company, to go for a public option in providing energy. I mean, it is unconscionable that the resiliency plan for California is shutting off the lights at various intervals as a way to prevent fires. The resiliency plan should become a movement that ends up with people owning the energy they need. That's what should happen in California."

"The worst thing that could happen," said the reverend, "is we think, after successive years, that running away from a fire is normal. It is never going to be normal. And a company making a profit from not solving a problem it should have solved years ago shouldn't be normal. Just like the levee breach in New Orleans after Katrina that flooded the city wasn't normal. All of this is the result of disinvestment in people and putting profit first. And that's what we need to solve."

A woman named Maria from Texas asked, "What

are the connections between the immigration detention centers on the border in the U.S. and climate change?"

"So here's a story," Saket replied. "About two or three years ago, doctors started noticing a substantial increase in kidney-related diseases in parts of Central America. They realized that because it was a lot hotter, people in agricultural work who are exposed to the heat became dehydrated and their kidneys started to feel the brunt of this. Conditions like that, climate change is responsible for. And people do what any normal person would do, which is that they try to do something to make a safer life for themselves and for their families. That is a very understandable, decent human response that runs up against a stone-coldhearted sort of government response like we're seeing in this country now, the detention centers. The same wealthiest investors in the world are profiting. The rise in the stock price of the corporations responsible for the management of the detention centers has skyrocketed over the last three years. Talk about ill-gotten gains or wages of sin. It's just appalling."

"Well, let's name names," I said, wanting to highlight the subject of our civil disobedience the next day. "BlackRock invests in the companies that are cutting down the Amazon rain forest. They invest in the companies that are cutting down the Indonesian forests in order to plant palm oil. They invest in companies that want to drill in the Arctic. They also invest in the immigrant detention centers, where kids are being held in cages. BlackRock is the name. If by any chance you have money invested in BlackRock, take it out."

In January 2020, the month after our teach-in on migration, public pressure combined with the decline in profitability of fossil fuel stocks led Larry Fink, the CEO of BlackRock, to announce that the firm would begin redirecting its investments away from the energy sector, primarily coal. In a **New Yorker** article, Bill McKibben, founder of 350.org, equated Fink's pledge with "cutting cake out of your diet but clinging to a slice of pie and a box of doughnuts."

The Rally

A large crowd gathered in Franklin Square the next day for the rally. There were representatives from almost every religious order and a Hollywood contingent as well. My friends Kyra Sedgwick, Maura Tierney, and Taylor Schilling had come to support and help me introduce the speakers. Oh, and the world-famous photographer Annie Leibovitz, decked out in a scruffy blue-knit cap, was there to photograph the event for **Vogue**, which was doing a story on me and Fire Drill Fridays. I know, odd bedfellows, but I thought it was cool that a magazine like **Vogue** wanted the story and Annie wanted to shoot it. I mean, if we need to go cultural . . .

Four days prior, the World Meteorological Organization released its state of the climate report to coincide with the opening of COP25, the gathering of world leaders in Madrid to focus on the climate crisis and the Paris Agreement. I began the rally by citing what the UN secretary-general said at the opening of the conference: " 'Climate change is no longer a long-term problem. We are confronted now with a global climate crisis. The point of no return is no longer over the horizon. It is in sight and hurtling toward us.'

"I wanted you to hear that quote because while we talk about specific aspects of the climate crisis here at Fire Drill Friday, it's important that we keep the big picture in mind. We need to understand that 100 per-cent of the climate scientists agree on this. There aren't two sides to the story. There's one side, the science."

There was an amazing roster of faith leaders at the rally: the Reverend Malik Saafir, a Methodist minis-ter, scholar, and community organizer; Liz Butler, a vice president of Friends of the Earth; the Reverend Kaji S. Douša, a senior pastor of the Park Avenue Christian Church in New York City and co-chair of the New Sanctuary Coalition of faith leaders who work to stop deportations; and the Reverend Noel Anderson, the grassroots coordinator for the Church World Service.

It was inspiring to see them all agree on the col-lective nature of the climate crisis and the need to pull together, particularly around the issue of climate refugees. Many of the speakers reflected Reverend Anderson's sense of urgency to call on policy makers and financial institutions to divest from detention

The Reverend Noel Anderson speaks.

centers and fossil fuel companies and switch to investing in "policies that reflect our values of justice, equity, compassion, and truth."

Another speaker was Imam Saffet Catovic, GreenFaith's senior Islamic adviser and a founding board member of the Global Muslim Climate Network.

"And I am an immigrant," the imam said. "I am the son of an immigrant and the grandson of an immigrant. The grandson of immigrants on my mother's side came through Ellis Island, and we know they did not have all their papers in order. Still America welcomed to her shores those who came from other nations, opening their hearts and hands because we know when immigrants are here, America does better.

Imam Saffet Abid Catovic speaks.

"On my father's side, I am the son of my late father Dr. Saffet, who came to this land as a political refugee. Yes, you can escape a land because you don't want to leave it but you have no choice. Those who come to the borders of our nation have no choice. They don't want to leave home, but they have no choice, and America has to honor the legacy of immigration because we are a land of immigrants."

Susan Gunn speaks.

Next Susan Gunn, director of the Maryknoll Office for Global Concerns, spoke on behalf of Catholic faith leaders she knew who were participating in climate strikes throughout the country at the same time she was addressing the crowd.

"I want to tell you the story of one mother and her children from Honduras who were traveling on a massive migrant caravan to the U.S. border last April. When she and the caravan passed through a small town in Guatemala where Maryknoll sister Dee Smith lives, Sister Dee was there to offer water and food. The mother immediately broke down in tears, exhausted. You see, one of her children uses a wheelchair. This mother was pushing her son's wheelchair from Honduras through Guatemala across Mexico to reach the U.S. border. It's not hard to guess why. In Honduras, the entrenched poverty and the worst criminal violence in the world have long made life difficult. Also, they have had drought in Honduras for the past four out of five years. That forced this mother and many people to flee their homes and caravan north.

"Yes, global temperatures matter, and what about people? We need the banks and financial institutions to stop funding and profiting from fossil fuel industries and delaying our shift from an economy based on fossil fuels to one based on efficient renewables. Together, my beautiful friends, we can lift our voices and fill the streets to make them hear us. To make them hear the cry of the earth and the cry of the poor by being here together today, we are already making the shift to a more sustainable civilization. Thank you."

Next we switched to younger speakers, and the stories that they told about being climate refugees were tough and emotional. Claudia Quinonez, an organizer for United We Dream and an immigrant, "undocumented, unafraid, unapologetic, and here to stay,"

Claudia Quinonez speaks.

described "being six, seven years old and my family not being able to buy food because there was no food. We didn't have water, and people killed over basic necessities." Joshua Álvarez, a member of the Sunrise Movement whose parents came to the United States from Mexico thirty years ago, enlisted in the Peace Corps when he finished college. He was interested in the reasons people now choose to flee to the United States, as his parents had. He was stationed in western Costa Rica, a region called Matambú.

"This is where I saw firsthand the impact the climate crisis is having on a global scale. My friend Marvin no longer wanted to live in Matambú because there were no work opportunities there because the traditional jobs like harvesting rice and corn did not pay enough.

"I noticed that the impacts of the climate crisis weren't just affecting Marvin. Throughout the two years that I was there, the crops were no longer producing the same yields due to unpredictable weather patterns. This caused many to migrate to the nation's capital, and some shared with me their desire to come to the U.S., just like the many migrants and climate refugees from Central America who are currently being held in concentration camps on our border.

"Let us acknowledge that the climate crisis is an intersectional issue that ties together racial, native, migrant, and even justice. In the words of Audre Lorde, 'There's no such thing as a single-issue struggle because we do not live single-issue lives.'"

Maura Tierney welcomed to the stage Rabbi David Shneyer, who read the new rabbinic statement on the climate crisis, Elijah's covenant between the generations to heal our endangered earth.

Rabbi David spoke: "I will send you the prophet Elijah to turn the hearts of the parents to children

Rabbi David Shneyer blows a shofar, a ram's horn trumpet.

and the hearts of children to parents. Our children and grandchildren face deep misery and death unless we act. They have turned their hearts toward us. Our hearts, our minds, our arms, our legs, are not yet turned toward them. Can we more fully turn our hearts to these our children? Join our young people in urging our governments to legislate a swift and massive program that intertwines ecological sanity and social justice as they were intertwined in the biblical practice of **shmita**, the sabbatical seventh year from Leviticus 25. Shaping all of these efforts as expressions of joyful community and not fearful drudgery."

Rabbi David picked up the shofar, the ram's horn, to draw attention to the urgency of the cause.

"In ancient times and throughout history, we've used the ram's horn, the shofar, to get our attention, a warning. Also, it's

Annie Leonard and Maddy Carretero hold hands as they serve as marshals from the march to the rally site to BlackRock.

a sound of liberation. So, first, the sound of warning. And we address this to everyone here and to all the leaders in this nation and to all the people in this world, the sound of warning, and then it will be followed by a sound of hope and liberation."

And the rabbi blew the shofar, and the ancient sound of warning and then of hope and liberation spread across the park as we formed ranks for the largest march yet.

There were maybe two thousand of us, and it was joyous. The Sunrise Movement and Fridays for Future each brought mammoth, inflatable globes. I caught site of Annie Leibovitz's blue cap above the heads of all the other photographers.

The environmental movement is quite diverse, which makes sense because black and brown people are the most impacted by pollution, oil wells, and severe weather. Many people associate environmentalism with the older, traditional environmental organizations made up largely of white men. To counter this misunderstanding, I always wanted to ensure that our speakers of color were prominently featured with me in the front of the march.

As we got closer to the office of BlackRock, our primary target of civil disobedience, word came that the company had locked its doors. Six activists were handcuffed to the doors and to each other in such a way that the police officers weren't able to saw the cuffs off them. The march organizers made a fast decision to detour to a Wells Fargo bank to hold our

A group of children participate in the march with homemade signs and an earth globe.

demonstration there. After all, Wells Fargo was deeply implicated in the climate crisis.

When we got to the bank, someone managed to get me through the crowd to the front, where many young climate activists were speaking and singing. I felt good sandwiched in among them, silver threads among the gold. I could see Reverend Fletcher in his red robes, and, sure enough, there was Annie's blue cap bobbing up and down as she searched for the most interesting shot.

The young activists gave speeches and chanted into a megaphone that they passed around. A young Asian man, clearly a leader, suddenly began singing, "We are a gentle, angry people / And we are singing, singing for our lives," and it was all I could do not to weep. My wonderful friend the activist singer Holly Near had often sung that song during our first Indochina Peace Campaign tour in 1972. In my heart, the song is associated with a time when I felt confident we could win: cut off funding that was keeping the Vietnam War going and get Nixon out of office because of the Watergate scandal. Here was Holly's song again, now part of the movement's great songbook. But I was less confident now, and the global stakes were even higher.

After a while, as the crowd in front of Wells Fargo dispersed, a large group of us marched over to lend our support to the handcuffed activists at BlackRock. A dozen or so police officers stood around, scratching their heads over what to do about the six people handcuffed to the front door. We chanted. I shouted our gratitude for the protesters' courage. And, figuring the police were just waiting them out, I took Kyra, Maura, Taylor, and my friend Emily Bickford to the Founding Farmers, a farm-to-table restaurant near my hotel that makes a great tropical drink.

I had just ordered my second drink when the restaurant owner came to our table.

"I never do this," he said. "I've had Justice Kennedy here, Ruth Bader Ginsburg, the Clintons. I never come out. But I had to come thank you for what you're doing. And the dinner's on the house." And with that he left.

Boy, did I feel swell. That night I slept twelve hours.

The group arrives at BlackRock to block the entrance, after the rally.

What Can I Do?

Many people don't realize the connections between the climate crisis and increased migration, so talk to your family, your neighbors, and your co-workers about what you have learned. Encourage them to share their families' stories of migration to the United States; remember that unless they are indigenous, all U.S. residents are immigrants. It's also important to realize, as Saket Soni reminded us, that "this issue is not [just] a faraway problem." Thousands of climate refugees are displaced within the United States, including, for example, all those forced to move because of Hurricane Katrina, Hurricane Sandy, and Hurricane Harvey and the tragic fires in California, among other extreme weather events amplified by the climate crisis.

As a society, we need more humane immigration policies, but we also need cultural change. We need to increase our empathy and overcome our fear of outsiders so that we can provide the refuge and support migrants need.

As Saket asked, "What if that climate refugee was you? What if it wasn't someone far away? What if it wasn't someone who was crossing an international border?" And whether they're from Texas or Syria, the United States must honor its history as a land that welcomes immigrants.

We can increase our understanding and empathy by getting to know recent immigrants—respectfully, of course, so as not to tokenize them. Invite them over for meals or to community events. Learn about their cultures, religion, traditions. You can work for environmental and cultural change at the same time— "feed two birds with one scone," as I heard someone say once. If you don't often encounter anyone different from you, cultivate a more diverse circle of friends and neighbors through volunteer or cultural activities. Our lives really are made richer, our perspectives broadened, and our communities strengthened by having diverse social networks.

While we're building that cultural change, we also

need to improve policies and practices around how immigrants are treated.

We can start by opposing the inhumane treatment in border detention centers. In 2019, the UN high commissioner for human rights, Michelle Bachelet, was shocked by what she saw at the immigrant detention centers on the U.S.-Mexico border. "In most of these cases," she said, "the migrants and refugees have embarked on perilous journeys with their children in search of protection and dignity and away from violence and hunger. When they finally believe they have arrived in safety, they may find themselves separated from their loved ones and locked in undignified conditions. This should never happen anywhere." She concluded that detaining migrant children may constitute cruel, inhuman, or degrading treatment that is prohibited under international law. So, everyone, please ask your elected officials to reunite families and ensure migrants are treated fairly and humanely—as we would want to be if forced to flee our homes.

Elsewhere in this book, I talk about divesting from firms and banks that fund fossil fuel projects. The same goes for immigrant detention centers. Research the companies invested in by your bank, 401(k) or pension fund, and other investments you have influence over to ensure they do not include any involved in immigrant detention centers.

There are many campaigns to improve U.S. immigration policy and practice. Good places to start are the youth-led United We Dream, which has tool kits for educators, for making your city a sanctuary city, and for supporting new arrivals. RAICES (the Refugee and Immigrant Center for Education and Legal Services) provides pro bono legal and social services for immigrants and has a children's program to help kids in detention. If that were my kid in need, I'd hope that kindly strangers would help; in this case, those kindly strangers are us. Migrant Justice, or Justicia Migrante, runs campaigns to support immigrant workers, stop deportations, and free immigrants from detention. The groups featured in our Thursday teach-ins and Fire Drill Fridays also lead campaigns for change and need support and volunteers: Green-Faith, Resilience Force, the New Sanctuary Coalition, the Church World Service, and Presente.org.

Sally Field, Winona LaDuke, and Clint Sobratti walk with Jane from the morning briefing to the Capitol, where the rally will be held.

CHAPTER TWELVE

Jobs and a Just Transition

I've talked about jobs and a just transition before, in the Green New Deal chapter, but the issue is of prime importance, and so, on December 12 and 13, we gave it a teach-in and Fire Drill Friday of its own.

When environmentalists supporting a Green New Deal talk about moving away from fossil fuels and into a clean, sustainable future, they always emphasize the concept of a just transition for the workers and communities whose lives will be impacted by that transition. But a lot of people don't understand what it means, or why it's important, so I want to be very concrete.

Those in the fossil fuel industry have been able to earn salaries in the range of $48,000 to—for those with unions—$125,000 a year with good benefits and pensions and collective bargaining. Thus far, most jobs in the renewable sector aren't unionized, and workers earn far less. Are we going to expect $125,000-a-year workers to be happy working for $25 an hour installing solar panels or energy-efficient windows with no benefits and no collective bargaining rights? No. We have to make sure that as workers transition from the industry that has allowed them to support a family, buy a home, and live in a decent community, they are able to continue earning comparable salaries, pensions, and benefits. The transition off fossil fuels is inevitable, but it being a just transition is not. That's up to us. If we commit to work together, the transition away from fossil fuels won't leave anyone behind and all workers and communities will come out the other side healthier, more resilient, and more prosperous.

If you want to know what the lack of a just transition looks like, take a look at coal miners. The mines are shutting down, their owners often declaring bankruptcy and refusing to pay their longtime workers back wages or health benefits. Many of those workers suffer from work-related black lung disease. Such callous disregard for the workforce results in the kind of desperation that led to the opioid crisis.

Think about the disruptive transitions brought about by globalization or the various trade agreements, or by the advent of intelligence technology. Displaced workers didn't end up with better salaries or pensions and benefits; they ended up with a lot of insecurity and anger. No wonder many others are worried about what a transition off fossil fuels means for them.

What has to happen, and what the Green New

Deal calls for, is a federally mandated commitment to involving unions, workers, and communities in decisions about transitioning to new energy sectors; to paying workers during training for new jobs that pay union wages and provide benefits and pensions; and to allowing workers too old to transition to new jobs to retire with dignity.

We mustn't fool ourselves into believing that just because a job is "green," it's good for workers. Renewable energy companies, if unregulated, are fully capable of operating under the same anti-worker, anti-community, racist, undemocratic framework as the fossil fuel industry. A solar company, Bright Power, fired workers after they started to organize a union. Tesla, the big electric car company and renewable energy darling, has repeatedly been found in violation of labor laws. In Mexico's Isthmus of Tehuantepec, two thousand wind turbines have been built, and three thousand more are planned. From a narrow environmental perspective this looks like a success, but not for the local community. In violation of their international rights, the local people were never consulted. Instead, the wind turbine companies ran roughshod over the territory's governing body and made contracts with small landowners and bribed local leaders. And, to make matters worse, nearly all of the electricity produced is supplied to foreign enterprises and private companies including the U.S.-based Walmart, not for the benefit of the local citizenry. Clearly, renewable energy companies are quite capable of repeating the same practices for which the fossil fuel industry has been criticized.

To avoid this, and guarantee a just transition, we must ensure that the renewables industry isn't concentrated in the hands of a few companies, as is starting to happen, but is decentralized. Already, five wind turbine manufacturers alone make up more than half of the global market share. Right now, the majority of abuse from the renewable energy sector is happening in Central and South America, Southeast Asia, East Africa, and the Democratic Republic of the Congo, where 60 percent of the cobalt that is used to build electrical batteries comes from and where it is mined under unethical conditions that include child labor.

In fact, renewable energy ranks as the third most dangerous sector for people defending their lands. Only mining and agribusiness are more dangerous.

Then there's Europe, where some renewable energy companies have started moving to countries where labor is cheaper, which means increasing numbers of their green employees are left jobless. Gonzalo Díez, secretary of the CCOO, the largest trade union in Spain, says, "Europe must put measures in place and demand a commitment from these companies that goes beyond the environmental." Otherwise, trade unions warn, the objective of a just transition included in the Paris Agreement on climate change will be put at risk.

We need to do this in the United States as well. We

need to hold energy firms accountable for their activities and the impact they're having on workers and local communities in the United States and around the world. According to the International Labour Organization, sustainable development encompasses three areas: economic, social, and environmental. A company that refers to itself as "sustainable" needs to comply with all three requirements.

As we create a sustainable economy, there are going to be millions of additional jobs beyond those in renewable energy. Workers will be needed to expand public transportation, upgrade public water systems, modernize the electric grid, build climate-resilient infrastructure, and rebuild after extreme weather events—all offering the potential of good union jobs that sustain both families and the planet.

If the United States sets the standard by enacting a Green New Deal, with its federal mandate that commits policy makers to a just transition, other countries will be more likely to follow suit. This would mean a national economic mobilization on the scale of the original New Deal back in the 1930s, one that tackles economic inequality and climate disruption at the same time.

It's encouraging that policy makers together with unions and environmentalists are starting to work in states and in cities all around the country with very concrete plans, rural and urban, that would create a just transition, even before the next election. The

Illinois Clean Jobs Coalition has included labor partners to help develop proposed legislation (the Clean
Energy Jobs Act of 2020). The plan is centered on
job creation and equity principles and is being structured to support workers and communities who will
face a transition from fossil fuel and nuclear power
generation. As of this writing, the legislation is still
being finalized in anticipation of votes by the Illinois
General Assembly in the spring of 2020.

In Colorado, a labor-environmental coalition organized and won legislation in 2019 establishing the Just
Transition Office in the state's Department of Labor
and Employment, charged with creating an equitable
plan for workers and communities transitioning away
from coal mines and fossil fuels.

In Maine, the state legislature passed and the governor signed a state-level "Green New Deal" bill—
designed to move Maine to a renewable energy
economy with the active participation of the state
labor movement in shaping the legislation.

The Teach-In

For that Thursday's teach-in, the room was filled with
young members of Fridays for Future, the student
movement inspired by Greta Thunberg. Some came
from New York, some from D.C. I had been able to
meet with them that afternoon and was impressed by

their fearlessness and smarts. Some of them had been climate striking every Friday for a year. They all had drawn wide-open eyes on the palms of their hands like the students who had traveled to the UN COP25 climate conference of world leaders in Madrid. Greta had just spoken there. The eyes were a signal from the world's youth: "COP, we're watching you. You better come out of there with strong bold actions and not just a lot of words." Unfortunately, the global climate conference ended without plans for strong action.

Each of my three guests for the teach-in brought a unique perspective to the issue of jobs and a just transition. Samantha Smith had just flown in from Norway, where the International Trade Union Confederation's Just Transition Centre, which she directs, is based. She represents the global union movement that is helping unions and their allies get concrete plans on how to actually advance the just transition.

Michael Leon Guerrero is the executive director of the Labor Network for Sustainability, has previously served as the national coordinator of the Climate Justice Alliance, and is deeply embedded in the domestic labor and climate movements.

My friend Winona LaDuke, co-founder of Honor the Earth with the Indigo Girls, is a rural development economist, growing industrial hemp on Ojibwe tribal land in northern Minnesota. She kept us focused on the importance of economic development that is region and ecosystem specific, including reviving this country's materials industry that used to thrive here before moving overseas.

I started by asking Samantha to give her vision of a just transition.

"In a just transition," Samantha began, "all jobs and especially green jobs are good jobs. It's not just the new jobs to which former fossil fuel workers might transition, but every job should be good. Every worker should have rights at work, like the right to form a union. Everyone should earn a decent living. Everyone should have health care. Everyone should have a dignified old age. And that is equally true for jobs in solar, if you're working in the Tesla factory or if you're building offshore wind farms.

"The other part of just transition is that some of these sectors with lots of emissions—fossil fuels, mining, power, steel, auto, transport—they are going to have to change massively, and some will be phased out. People's jobs in these sectors will change, a lot. Just transition for workers in these sectors means different things for different groups of workers. Older workers should get a bridge to pension. Younger workers should have retraining and reskilling, and a path to a new and decent job with comparable wages. Everyone should have what in Europe we call social protection, which means health care and income support. And no one should be left behind.

"The last thing about just transition is that workers live in communities. And so a big part of just transition is regional redevelopment so that workers in sectors that get phased out can get a new good job where they live. People don't have to be mobile; they can stay where they are. And they can also help to build the new economy, the low emissions economy that we need."

Michael joined in. "First of all, thank you so much for taking on this really important topic. I honestly think it's probably the most important movement challenge that we have. So there is this whole perceived dichotomy between jobs and the environment, which is really a false dichotomy. But there are very real tensions between sectors of the labor movement and the climate movement. If we don't bridge that divide, honestly, I think we can't really win the policies that we need, like a Green New Deal. And I honestly don't think we can win a Green New Deal without the U.S. labor movement. We're talking about twelve million union members in this country and their families. That is a huge social force. That's going to be critical for this. And Franklin Roosevelt, with the original New Deal,

"Everyone should have what in Europe we call social protection, which means health care and income support. And no one should be left behind."

understood this. The New Deal was not so much a social welfare program as an empowerment program. Besides creating eight million jobs, FDR and Congress also passed the National Labor Relations Act and other pro-labor things. Because what FDR understood was that to win the New Deal, to implement it, and to defend it, he had to turn it over and put it in the hands of the people. And they had to organize. And that's what they did, and the labor movement just exploded in the era of the New Deal. And not only that, but income inequality was at its lowest in the wake of the New Deal policies that were passed.

"Since then, it's gone in the opposite direction. The gap now, between the wealthiest and the rest of us, is almost higher than it's ever been, right? And capital has done everything it could in that period to roll back all of those gains and to redistribute wealth to the top. There was a report that just came out, for instance, by the Economic Policy Institute that talked about how $340 million is spent a year, by employers, to fight unionizing campaigns in this country. Over 40 percent of those campaigns were found in violation of labor laws. So that's what we're up against. And it's no accident, for instance, that in the last fifty years, ever since Lyndon Johnson gave an address before Congress, recognizing that climate was going to be affected by carbon emissions, that in that same period, income inequality has increased at the same rate that carbon emissions increased. Here's a graph that shows that comparison. So the Green New Deal

TOP 1% SHARE OF INCOME VS. CO$_2$ EMISSIONS

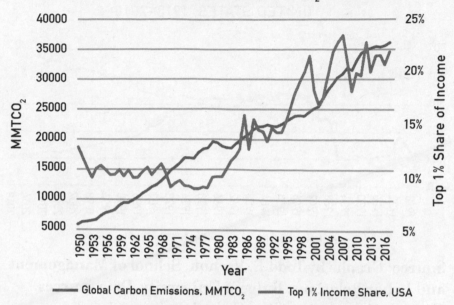

Source: Graphs by Todd E. Vachon, School of Management and Labor Relations, Rutgers University. Data sources: Historical Statistics of the United States, unionstats. com, Saez (2020); Global Carbon Project: Carbon Dioxide Information Analysis Centre.

has to empower workers and communities to implement it, fight for it, defend it, and build it.

"There's no pathway to get to the economy that we need to protect our climate without going to a just transition. And the unfortunate thing is, in the U.S., we probably have had more bad examples of transitions than good ones.

"We talked to workers that haul coal in Pennsylvania. They said, 'Look, we've been through this three times now. Each time they laid off five hundred workers. We went through the training programs and all that,

UNION MEMBERSHIP AND TOP 10% SHARE OF INCOME, UNITED STATES, 1918–2018

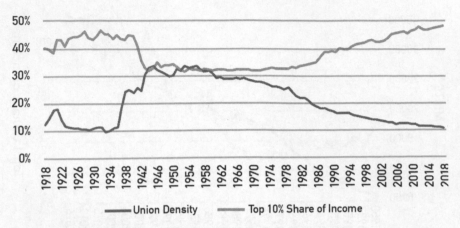

Source: Graphs by Todd E. Vachon, School of Management and Labor Relations, Rutgers University. Data sources: Historical Statistics of the United States, unionstats. com, Saez (2020); Global Carbon Project: Carbon Dioxide Information Analysis Centre.

and then there were no jobs, or the jobs honestly just sucked.' And we hear this story over and over again. So workers, the labor movement, needs to be at the table and so do communities in shaping not only policies but the implementation of those policies. It has to be in the hands of the communities and workers, because communities aren't identical. There aren't cookie-cutter solutions."

"That's right," said Winona. "One size doesn't fit all. The nature of this world is ecosystem based, so a Green New Deal or any green economy means different things in different places according to what there is and what people need. At some level, I believe that things are going to have to be far more localized. We've got to move less stuff around and get more efficient. And I think that we need to build regional economies."

"We can be creative with the transition," said Michael. "It doesn't have to be the same kind of model everywhere. There are a lot of solutions that exist in communities. And if we empower people to do that and communities where our solutions are already being developed, there's tremendous potential, and it even plays out in industrial things. I was talking to a bus mechanic in San Jose, California. And he said, 'We went from diesel buses to hybrid buses and now we're going to electric buses all in a span of ten years. When we got the hybrid buses, they gave us a fleet of five of them, and they didn't give us any parts. So we kept one bus just to strip it for parts. And then I had to train the workers on how to do it.' So that's just another example that if we don't actually have the workers, the communities, that know the stuff from the ground up involved in it, we're going to do it wrong, which is why it has to be something that's owned by the workers.

"I'd like to say something about democracy and how you get people on board with transition. Because people have to agree to transition in democracies. They must consent to change.

"Let's use Germany's agreement on a just transition

in the power sector and coal mining as an example. Germany is a big economy. It burns a lot of coal. They want to do things on climate. The unions understood this and were actually pushing the government to set up a commission to tackle the political problem of how to phase out coal-fired power, which employs a lot of people. There was political pressure from the environmental organizations, too, so finally the government set up a dialogue with employers and unions, local government, the federal government. And this year they reached an agreement that will completely phase out coal-fired power in Germany by 2038, if not before, along with a just transition for every person who today is employed in coal-fired power and coal mining.

"How do you get coal miners in Germany to support a phaseout of coal mining? But that has happened, right? There's somewhere between twenty thousand and thirty thousand workers who today work in coal mines and coal-fired power and whose unions negotiated a just transition and phaseout of their sectors.

"Getting to Germany's just transition agreement was a long process. What was key for coal workers was that no worker will be left behind. No one will be fired. The older workers will be taken care of. They get a bridge to pension. They and everyone else have health care. The young workers will be retrained while they're working. They won't have to pay for their retraining. They'll be placed in a new good job. If they go into a job where they're earning less, they'll get a lump sum that will bring them to the salary level where they were when they worked in coal.

"So today's German coal miners will have other, good jobs as a result of this agreement. Many of them will still be working in energy, but it's going to be clean energy. And the regions where they live, including the regions in the former East Germany that were deindustrialized after the reunification of Germany, will get €40 billion over twenty years from the federal government for green infrastructure like high-speed rail and broadband, soft infrastructure like schools and hospitals, and other things that create jobs and make it good to live and stay in the regions.

"We see this kind of process a lot in the Just Transition Centre. We work basically only with workers and unions in high-emissions sectors, coal power, nuclear, mining, autos, steel, transport, construction. Workers and unions from these sectors are like anyone else, they understand that climate is an issue and change is coming. They just lack real pathways to new low-emissions jobs that are also good. They will negotiate a just transition and agree to it when the new jobs are as good as or better than the jobs workers have today.

"A just transition requires coming together with labor at the table, with communities, employers, and government at the table. That process of negotiating just transition means that workers shape their own futures. But that future has to be concrete if people are going to trust it. If you're an industrial worker, you probably want to continue to be an industrial worker. You'd be totally happy to get more skills, a better job, ideally better pay and new prospects in the green economy, but you have to see the steps and commitments for that to happen in a concrete way."

"One more thing about Germany," Samantha said. "The federal government is important. It's backing up the agreements with the workers and regions. The government, parliament, has passed all the different laws and regulations and budgets that are needed to make this agreement happen. But maybe the most important thing from my point of view is that the mining and energy unions were at the negotiating table throughout this whole process, even though the process was about phasing out some of their members' jobs. As soon as they reached an agreement, they sprinted out of the negotiating room to go and socialize this agreement with the members. So they were committed to this agreement, right, because they knew that this is the best possible deal they're going to get for their members and also for the regions where their members live. It's a very, very good deal for the people in the communities and for the people working in that sector."

I asked Michael if he wanted to add anything. "I want to mention a domestic example," Michael said. "The State of Colorado has established a Just Transition Office, the first of its kind in the country, and it's really significant. It's focused right now primarily on the coal regions in Colorado, but it's significant because it's one of the first initiatives that is actually thinking ahead of this problem of just transition. Most of the reasons why transition programs in this country have not been good, for the most part, is because it happens at the tail end. A plant shuts down or we go through NAFTA, and people lose their jobs, and then it's not thought through well. The idea of a just transition office, actually planning ahead about how we're going to deal with transitions that are already happening in the economy, is something that we need to look at, at the state levels and also the national level. That foresight is going to be critical for us going forward."

Our livestream audience for the teach-in was just like me, full of questions! Naomi from Washington, D.C., wanted to know how the Green New Deal would be more equitable than the original New Deal.

Samantha answered, "Well, you could learn something from the apprenticeship programs that the building trades have now, which is that you take a certain number of apprenticeship jobs for women, for people from communities of color, for people who today are excluded from participating in the economy. So, for example, in Canada, the apprenticeships are often for new Canadians, for immigrants. That can be one way to go if you're going to have a massive infrastructure building program. You set aside training programs, so people are getting skills on the job.

And then they will be certified in their trades and can go on to have other jobs using those skills. It's very effective."

Eric from California wanted to know if unions are in support of the Green New Deal.

Michael answered, "As you know, we've got a number of Green New Deals, right? But I think labor is looking at them in terms of how workers' rights are going to be protected in any of these Green New Deal policies.

"There are a number of resolutions that were passed by central labor councils and some unions. The strongest one probably was passed by the Service Employees International Union, one of the first international unions to pass it. And we had the Association of Flight Attendants, the Communications Workers of America, the United Electrical Workers, and then a number of central labor councils and state federations. The language that we're seeing from a number of them, though, is supporting a Green New Deal but not necessarily the resolution that was introduced by Alexandria Ocasio-Cortez and Senator Edward Markey."

Samantha added, "Many labor unions wouldn't mind a Green New Deal, but they want to be part of shaping it. At the moment the Green New Deal is a resolution. It's at the principle level. It's easy to misunderstand each other when it's conceptual and not specific. On the other hand, when you can talk about concrete plans for good, new jobs, for health care and pension, that's something people want. If you ask workers, 'Okay, how do you feel about building schools or roads or offshore wind turbines or something else with the same wages and conditions you have today or better?' They would be for that."

Winona agreed, saying, "For instance, there is hydrogen fluoride at these refineries, and workers don't necessarily want to work in a really dangerous situation. And so if you have a worker at a refinery like the Husky refinery in Superior, Wisconsin, and they're faced with the fact that they may blow up any day and you say, 'How about we transition out of this one and come up with something less toxic?' You bet most people would sign up. They want the security of a job, but they aren't necessarily wedded to hydrogen fluoride."

Wendy from Nebraska wanted to know if there are successful examples of people switching from the fossil fuel sector jobs to green economy jobs.

"Yes," said Samantha. "There are examples in this country, and there are examples in other countries. I live in Norway, where about 60 percent of the output of our GDP is coming from state-owned companies. In Norway our big oil and energy company is actually putting a lot of its resources now into offshore wind. And in Denmark there was a state-owned utility company that's gotten rid of all its coal and gas assets and is now the world's biggest offshore wind company. They're going to build a ton of offshore wind turbines off the East Coast of the United States. And we also have examples even at a city level where people go from building one kind of pipe to another kind of pipe, from working in a fossil-fuel-driven transport system to working with electrical buses. There's a lot of examples happening all the time. The thing is, it has to happen really fast, be scaled up really fast, and we have to be sure no one is left behind."

Frank from Ohio wanted to know what types of low-carbon jobs people could be trained to do.

Michael answered, "The International Brotherhood of Electrical Workers in California, for instance, and in New York, are developing really good apprenticeship programs. And Massachusetts is also. In California, you can get a solar installation career at $75,000 a year, right? I think the infrastructure that unions have for training workers is a tremendous resource, whether it's pipe fitting, whether it's electrical work. And in the end of the training there's a real job. Charging stations will be needed everywhere for what will be a national fleet of electric cars, buses, and trains. In fact, a really good example in Los Angeles is a campaign called Jobs to Move America. It's doing a really tremendous job making sure that the manufacturing of trains and buses happens here in the United States for electric vehicles. The United Steelworkers Local 675 in Los Angeles represents refinery workers in Southern California, and they just organized a Proterra plant, which is an electric bus manufacturing facility. They are very clear that the future is not with these fossil fuel jobs. They're looking at what are the other industries where we need to be organized and where we need our workers to get involved."

The Rally

The next day's Fire Drill happened in 30-degree cold and pouring rain, yet the crowd was big and engaged. I enjoyed seeing petite Sally Field marching to the rally next to statuesque Winona, towering more than a foot above her. They were so deep in conversation that they kept falling behind, and I'd have to send people to bring them up to the front where the cameras were. We stood at the podium with a canopy of umbrellas over us, held up by other marchers. All around the stage were young people from Washington, D.C., and Baltimore who came to march and speak about the climate crisis problems they would face when they became adults.

The first speaker, Kristin Taylor, who has taught for thirteen years at Hamilton Elementary/Middle School in the Baltimore City Public School System, was joined by many of her seventh graders.

Kristin wanted to call attention to the students in the crowd first: "These kids standing here behind me, and sitting in many classrooms around the country, are why we have a special responsibility to confront the reality that our climate is in crisis. They do not deserve to be tasked with solving a problem a previous generation created, but they are willing to step up

Kristin Taylor speaks.

and try to stop the damage. It is not just your house. This is their house, too. The clock is running out, and we need to demand no more business as usual. We demand action now.

"The climate crisis needs a bold, ambitious, and urgent response. As we continue to do our part in our jobs, we need others in their jobs to develop policy details. Our job is to educate and empower our students. When our country was faced with a massive economic crisis ninety years ago, the response was a massive investment to create jobs and save the communities, the New Deal. As a history teacher, it's difficult to ignore the pattern of history repeating itself. I know that our country and entire planet face the worst crisis in history. It is both a crisis of the climate and a crisis of growing inequality. We cannot address one without the other, which is the idea behind the Green New Deal. By transitioning to renewable energy while using cleaner business practices as a whole, not only will our planet be given a chance, but so will our economy.

"I am here as a public school teacher, but I'm also here as a worker. All of us in the school community—teachers, support staff, nurses, and administrators—are all workers, community members, and parents who greatly care about the climate crisis and the future of jobs and the environment for our kids and grandkids. Our jobs in the schools are more critical than ever before because we are responsible to develop children to become critical thinkers, active members of our communities, and responsible young adults, to be productive citizens of our society."

Tanaya, a member of the Hamilton Elementary/Middle School Social Justice Club, added, "As one of the newest generations, it's important that our teachers and community members hear us and allow us to speak out, too. As children, when we grow up, we want to do extraordinary things. But these dreams may be harder to achieve because of our environment. Policy makers need to hear us because our futures depend on you picking up your pen and making the needed transition to renewable resources."

Sally Field spoke after the students, addressing the crowd as a grandmother but also as the woman who won an Oscar for the labor classic **Norma Rae**, the story of a woman who unionized her textile mill.

Baltimore school students Tanaya and Ruby speak. They came with Kristin Taylor.

"I was lucky enough to star in a film called **Norma Rae**. Based on a real hero, Crystal Lee Sutton, a fighter, a unionizer who fought with her community of black and brown and white workers to change her workplace in the textile mills, the textile industry. And she changed it, she did it. But when those industries had to pay them higher wages, a slightly more than living wage, and better working conditions, they left and went to another country altogether where they could pay the workers next to nothing and treat them more abominably because the executives wanted more money in their pockets. They decimated Norma Rae's, Crystal Lee Sutton's, communities because they had no just transition.

"That's what they're here to talk to you about today. Where people can transition to a better job, a greener job, in a workplace that will support them and their families with more than just a living wage. This is actually happening. It's not pie in the sky. It's not a dream over the rainbow. It's real. But what has to happen is everyone has to get up out of their comfort zone and scream now." Sally pointed to the Capitol behind us.

Sally Field speaks.

"If those people don't listen because they're all fighting other battles against each other—if they're not listening—if we can't get them to listen and move, then do it yourself. Thank you for being here. I'm soaking wet, I'm freezing cold, and I'm proud as hell to be here. Thank you."

Clint Sobratti, a public bus driver in Montgomery County, Maryland, and a member of the UFCW Local 1994, spoke next. He is an elected chief steward, a member of the local bargaining committee, one of the founding members of the local green committee, and a member of the climate committee group in his union.

"Thank you for focusing today's event on working people and communities. American workers want to be the soldiers in the fight for transformation to clean and renewable energy, but we don't want to be left behind on the battlefield, wounded and struggling to survive. In my union, we're in the process of formulating demands that will help address the climate issue and directly help our members. We are bargaining for ourselves, our families, but also for the common good. But let's face it, this is a crisis so huge it can't be solved one county at a time.

"Today, I want to talk about how being a bus driver has taught me about justice. When I began this job, I realized I was no longer driving for myself. I had the responsibility for the passengers on the bus, and for

Clint Sobratti speaks as students from Baltimore join him onstage.

pedestrians, the cyclists, and other vehicles. This is a responsibility that my co-workers and I take very, very seriously. We hold human lives in our hand each and every day. Our buses are welcoming places for immigrants, people of color, young and old, physically challenged, rich and poor. We drivers make sure that all passengers are safe and respected. This is what democracy looks like. Our public transit system makes life better for many people.

"Imagine if there were fully funded public bus and rail systems all over the country, how much more of a difference will it make? I also know that we're playing a role in saving the planet. Probably most of us drivers don't think of ourselves as environmentalists, but we are. When we take the bus—when you take the bus instead of driving alone—you're helping reduce emissions. The faster we replace our fleet of diesel buses with zero emissions, the more we will help our planet. I support the transition to zero-emission buses, but I also want to say that equity and justice includes ensuring that all workers are protected. The technology changes every day. That means the workers who are maintaining and operating the vehicles need support and training also. Transition creates winners, but it also can create victims. As we work for a healthy and equitable community, we must not leave anyone behind."

Joe Uehlein, spoke next. He is the founding president of the Labor Network for Sustainability and Voices for a Sustainable Future and spent more than

Joe Uehlein speaks.

thirty years organizing, bargaining, and doing strategic campaign work in the labor movement.

"We are, all of us, climate warriors, labor and climate activists, working together to solve this problem. If we're not all at the table, we're not going to win. We need to get to that table and make sure that the Green New Deal is fair and just for everyone. For the working people, for people in communities, and for those who might lose their jobs in communities that might be damaged. For us at the Labor Network for Sustainability, that means five years of wage and benefit parity for impacted workers and block grants to communities so they can reorganize their economies around the concept of full-spectrum justice. That's what we're fighting for. That's what we all have to fight for.

"Now, the people in the building behind us," Joe pointed to the Capitol, "and those before them, they have failed us. They're not stepping up to the plate in the way that they need to. In the words of the late, great Robert Hunter, 'I don't know, but I've been told, if the horse don't pull, you've got to carry the load. I don't know whose back's that strong. Maybe find out

Sally Field gets arrested on the Capitol steps.

before too long.' Well, the reckoning is here, and we're finding out whose back is that strong. All of us, we know that our back is that strong.

"Climate change is the real job killer, not the solutions to it. It is decimating our public budget, which steals bargaining power away from unions that negotiate in the public sector. We have to understand that we will only win if we do it together, labor and climate activists. We've had our differences, but we need to put them aside and come together and develop a strategy to make climate workers a real force. Now the basic question to me—and I'll close with this—raised before us here today is whether the government, which, of course, means the people, has the compassionate responsibility to see to it that we preserve a life-sustaining climate for current and future generations. Thanks for coming out today. Keep it going. Rock on."

And with that, we marched to the Capitol steps, and I watched as all these workers, teachers, bus drivers, labor organizers, and climate activists chanted and brandished their banners and, one by one, were arrested. Sally Field was right there with them.

Watching the camaraderie and solidarity of them all risking arrest together gave me a glimpse of the power the climate movement will have as labor and climate activists increasingly get to know and understand each other and realize how interconnected our fates are.

What Can I Do?

The phony choice between good jobs and a healthy environment has kept labor and climate activists divided for too long, so the first thing we all can do is to listen, learn, and stand in solidarity together. The better we know each other, the more we can work together for solutions that leave workers, communities, and the planet better off.

If you're in a union, start a climate committee to learn about climate issues together and build support for climate action. Get your union to pass a climate resolution calling for bold climate action, in line with the science. The Labor Network for Sustainability has great tools to advance climate education and policies in unions that you can download from its website, including a sample draft "Resolution to Encourage Labor Support to Help Shape the Green New Deal." Remind people of Joe Uehlein's point at Fire Drill Friday that climate change, not the solutions to it, is the real job killer.

If you're not in a union, there is still plenty of good organizing you can do. Talk with your co-workers about how the climate crisis is affecting your jobs and communities. Start a climate committee to work on climate in ways that make sense for your business. Every workplace, from an office to a hospital to a factory to a restaurant, can help raise awareness and drive climate solutions.

Reach out to union members in your family, neighborhood, or place of worship to ask what their union needs to get more involved in climate. As Michael Leon Guerrero of the Labor Network for Sustainability warned us during the teach-in, we can't win a

Green New Deal without the U.S. labor movement. Let's keep building bridges so we become one unified movement for a better future.

If you employ others, ensure your company's policies are not just pro-climate but also pro-worker and pro-union.

The solutions to the climate crisis and to the socioeconomic inequality crisis are intertwined. The more we have workers at the table from the start, the more successful our solutions will be. To help understand the impacts of climate policies on workers and communities, check out the great resources that the Labor Network for Sustainability has available.

And use your political voice! We can't paper over the fact that people employed by coal, oil, and natural gas stand to lose their jobs and livelihoods as we switch to renewable energy. Contact your members of Congress to demand that the massive subsidies given to fossil fuel projects be redeployed to provide comparable jobs—good union jobs—in clean renewable energy and climate adaptation. It's hard for renewable energy to compete with fossil fuels when our government gives billions of dollars of public money to subsidize them; that money should be instead supporting a just transition to a clean energy economy. Demand that Congress take care of these workers, providing safer, better jobs and fairer working conditions by supporting the Green New Deal. You can also advocate for Green New Deal–type laws at the state level, looking to Colorado, Maine, and New

York as potential models. If the local, state, or federal government complains about the cost, remind them of the enormous cost of climate inaction, or the military budget.

Above all, remember that worker rights are human rights. Justice for workers is fundamental to building a renewable energy economy that leaves no worker behind. As the brilliant young Tanaya told us at our Fire Drill Friday, "As children, when we grow up, we want to do extraordinary things. But these dreams may be harder to achieve because of our environment. Policy makers need to hear us because our futures depend on you picking up your pen and making the needed transition to renewable resources."

The scene inside the Hart Senate Office Building.

CHAPTER THIRTEEN

Health and Climate Change

As I was finishing this chapter, Trump announced a national emergency because of the coronavirus. Already **Grace and Frankie** had shut down production, supermarkets were facing shortages, and across the country people, including many who live paycheck to paycheck, were being told not to come to work. Hospitals were dealing with a severe lack of beds, ventilators, and protective gear for health-care workers.

Though our country was warned by the World Health Organization that this was coming, the U.S. government was slow to act in almost every way. Most critically, we lacked the ability to test early on, which would have helped identify where and how to slow down the virus's spread. You might wonder why a wealthy country like ours doesn't have the ability to immediately focus on identifying and responding rapidly to infectious threats. We actually had a government team—the Global Health Security and Biodefense unit—that the Obama administration created to ensure pandemic preparedness. It was dismantled by the Trump administration in 2018.

Why am I writing about this in a book about the climate crisis? Because this is a chapter about how the climate crisis is impacting health—right now, as you are reading this—and what needs to happen to prepare for and mitigate the worst of it. And just about every single thing that this government of ours has done is exactly the opposite of what must happen— with a pandemic or with the climate crisis: denial, lack of preparedness, disregard for science, viewing the federal government as merely a "backup." This is not a drill! It is a vital teachable moment.

As with the coronavirus pandemic, the government has chosen to deny the climate crisis or, for those who admit it, delay an appropriate response. This isn't simply a problem of truth telling. We are vastly unprepared for what is already happening (air pollution, cancer epidemics that could be prevented, extreme weather events, refugees, diminishing supply of clean water, diminishing access to healthy food, possibly even diminishing oxygen), much less what will be coming at us year after year and reaching a tipping point in a decade if we don't act fast.

As with the pandemic, scientists have been warning us in ever more alarming and unanimous ways about what the climate crisis holds in store even if we immediately start doing everything we need to do to avoid it . . . because we didn't start soon enough. Sound familiar?

As with the pandemic, to have even a slight chance of mitigating the effects of the climate crisis, we need a robust federal government.

If you ever thought that a strong central government wasn't needed, think about how different our country's response to the coronavirus would have been had the appropriate agencies been in place (and funded) and the needed testing been available. South Korea's expansive and well-organized testing, combined with immediate, extensive efforts to isolate infected people and trace and quarantine their contacts, helped that country's coronavirus cases to drop. That could have been the case in the United States had our federal government acted immediately.

And why weren't these things in place and available? Because we have a president and an administration who want to make the government so small it cannot interfere with the goals of corporate America. Never mind the poor, the old, the sick, the workers, the climate. That is why, at his March 13, 2020, Rose Garden press conference, Trump surrounded himself with CEOs from corporate America offering up ways they would help solve the health crisis. Trump has deliberately weakened the federal government, firing and not replacing professional staffers in major departments and turning instead to the private sector for help. We've learned from past experience that in a time of crisis that approach does not work.

On its own, the private sector can't move quickly enough to meet the need unless the government orders it to do so, provides the capital to change its production line to ventilators, harnesses the supply chain to provide parts and raw material, and guarantees it will pay for all the ventilators, or N95 masks, the company produces. The private sector fills the need, but the government must guarantee the financial support for this dramatic change.

As I have said a number of times already in this book, we face a collective crisis that demands a collective response. In November, we can help determine whether we elect leaders who will be prepared to face what's coming. Let the coronavirus be a teachable moment.

The week leading up to the December 20 Fire Drill Friday was intense. I was turning eighty-two on the twenty-first, and many friends were coming to support Fire Drill Friday and celebrate my birthday. We hoped that eighty-two people would sign up to risk arrest, but by Wednesday it was already clear there would be many more. I was excited and nervous . . . and tired. Teeth grinding and bed tossing, while engaged in furious imaginary conversations with elected officials, once again made sleep an issue as I learned more about how the worsening climate crisis was affecting people's health.

Especially children. Especially babies in utero! Why aren't pro-life people standing up against fossil fuels?

By that point, I had been arrested four times. I had, at most, one more arrest before I risked getting locked up for months. I had been saving my last arrest in D.C. for this Friday, figuring they would have to lock me up, which would mean I'd turn eighty-two while in jail. I knew that would get lots of press coverage. The purpose of Fire Drill Fridays, in addition to education, is to get the word out to as many people as possible that a crisis is upon us and we need to step into a new, brave frame of mind. Being willing to engage in civil disobedience is a part of communicating that sense of urgency. And this time, I knew the jail drill: lots of layers to soften the metal slab.

The Teach-In

One reason I had wanted to focus a Fire Drill Friday on health was that not enough people make the connection between the health crisis and the climate crisis. They don't link the climate crisis to the new strains of diseases, the increase of respiratory diseases, cancers, and other illnesses that strain health-care systems worldwide, including here in the United States. Global warming expands the geographic range of mosquitoes and ticks and rodents, and other carriers of diseases, bringing dengue, encephalitis, yellow fever, and Zika to places where they were unknown before. Flooding causes upsurges of rodent-borne illnesses, contaminating food and water, an impact on human health that continues long after the waters have receded.

Air pollution, intensified by global warming, has led to two million deaths worldwide from lung cancer, heart disease, asthma, and allergies. The World Health Organization estimates that every day 93 percent of the world's children under the age of fifteen breathe air so polluted that it puts their health and development at risk. That's just hard to wrap your head around. The health-care systems in this country, and around the world, aren't equipped to handle what already exists, much less what's coming.

Dr. Sandra Steingraber, the scientist with the heart of a poet who had been my cellmate at the very first Fire Drill Friday, was the first guest to speak.

"It's not often I'm introduced as both a poet and a biologist," she began. "And what I say about that is biology and poetry are both about the mystery of being alive. Whereas biology wants to solve the mystery, poetry simply says 'behold.' I became a cancer patient as a young woman. I was diagnosed right around my twentieth birthday with bladder cancer, a kind of cancer that is considered a quintessential environmental cancer. So, while my question '**Why me?**' required a biological answer, I needed poetry, too. Specifically, I needed women's poetry. And when I see young people at these climate strikes carrying signs that say, 'You will die of old age and I will die of climate change,' I feel like I know them. I know what it's like to be a young woman whose future is in jeopardy; to feel a kind of dread about not being able to protect myself, or lay plans, or know what I should do with my life, because a massive health problem lies right in front of me.

"As a biologist, I'm dedicated to looking at the intersection of climate and health with all the objectivity that we scientists bring to bear when we look at data. Scientific objectivity, for me, is not the same as political neutrality. At the end of the analysis, if what the science on climate and health shows us is that people will die, that people will be harmed, that there will be excess strokes and heart attacks and preterm births, that crops will fail and people will risk mass starvation and become refugees, then, as a scientist, I take this data and embed it in a social movement that can save lives.

"We all exist in our bodies in a kind of exquisite communion with the environment around us. We are 65 percent water by weight, which is to say that our blood plasma, our tears, our breast milk are raindrops, are groundwater. So we exist as part of the water cycle. With every breath we take, we are inhaling a pint of atmosphere. And 20 percent of that air that we breathe in is oxygen. And it comes from two places. Half of it comes from trees, and the other half comes from plankton in the ocean. That's the only source of oxygen that we have on this planet.

"The health crisis begins right when the drill goes into the bedrock. One drill goes into the bedrock to exhume fossil fuels, which come in three forms, right? Coal, oil, and natural gas, a kind of unholy trinity of fossil fuels. Air pollution is just another name for fossil fuel combustion by-products. One hundred thousand Americans killed by air pollution each year is 274 funerals every single day. There have been eighteen hundred studies showing how the extraction process of fossil fuels called fracking, which is the dominant way that we blast these things out of the ground, affects our health. These studies all show that people of color and low-income communities are hit first and worst. And that means [they suffer] things like birth defects, childhood leukemia, heart attacks, strokes, preterm births, making fracking the leading cause of disability in this nation.

"A little community in Massachusetts that I'm working closely with now, Weymouth, is a low-income immigrant community, already overburdened by toxic industry. So along comes an oil and gas company called Enbridge which wants to push gas from the fracking fields of Appalachia all the way into Canada by building a massive compressor station on a little piece of land near where three thousand children go to school, right next to low-income housing where immigrants live. This would damage air quality in such a way that it would almost certainly harm health, and yet it's going forward anyway.

"The good people of Weymouth are rising up and committing civil disobedience to try to stop this. There are Weymouth, Massachusetts, communities

The Reverend William Barber, Ai-jen Poo,
Katharine Hayhoe, Sandra Steingraber, and
Jane. Carla Aronsohn is filming them.

all over the United States where the people who have
the least resources are being asked to pay a terrible
price with their health to service infrastructure. As a
biologist at Ithaca College, I am interested in taking
action on the data that we already have. My current
work now is to try to launch a center for climate justice at Ithaca College where we can really bring science and activism together, because I think they're a
powerful combination."

Next, I introduced my friend Ai-jen Poo, a leading
voice in the women's movement, who is the executive
director of the National Domestic Workers Alliance
and co-director of Caring Across Generations.

Ai-jen began with our history as friends. "The
last time I saw Jane, she was walking the halls of
Congress with me and about two hundred domestic
workers advocating for a federal domestic workers bill

of rights. And that is who I represent. I'm here, representing the nannies who take care of our children, the house cleaners who help maintain order and sanity in our chaotic lives, and the home health-care workers who take care of our loved ones with disabilities and allow them to live independently in their homes, and they take care of our growing aging population in this country.

"These people are the first first responders in the context of climate disasters. They are there for the elderly and for people with disabilities. When the electricity goes out and there's no running water, they stay, even though they themselves have families in communities that are hardest hit by some of these crises, the last to be evacuated, and they're managing all of it. The average annual income of a home health-care worker is $15,000 per year."

Ai-jen Poo speaks.

FAR-REACHING HEALTH EFFECTS

Katharine Hayhoe

Katharine Hayhoe is an evangelical
Christian, an atmospheric scientist,
and a professor of political science at
Texas Tech University, where she's
director of the Climate Center.
As a climate scientist, she studies
every aspect of the planet, from the
far reaches of Greenland and
Antarctica to what's happening right
where we live. In 2014, she was listed
as one of **Time** magazine's 100 Most
Influential People.

"We often think of climate change
as an environmental issue,
but it's much more than that.
There are impacts on human health, especially that of the unborn, the babies, the
children, but there's at least six more ways
that climate change affects our health. First
of all, it affects our health through making
heat waves stronger and more intense. This
summer around the world, we broke more

than four hundred all-time heat records, and Australia just had the hottest day across the whole country. That affects us, especially the elderly, the children, and people who are already suffering from lung or heart disease.

"The European heat wave of 2003, which was twice as likely to occur as a result of a changing climate, was responsible for more than seventy thousand deaths. We're seeing stronger hurricanes, more frequent flooding, rising sea levels, and these directly affect our health when we're exposed to them, and they can lead to things like long-term contamination. When hazardous waste sites in Houston were flooded by Hurricane Harvey, where it's estimated that 40 percent of the rain that fell would not have occurred if the same hurricane had happened a hundred years ago, those hazardous waste sites were flooded. Where did that water go? Into the poor neighborhoods.

"Then we have its impact on our mental health, our anxiety, which I see growing and building over concern about what our future holds, and even PTSD after people have been exposed to these increasingly devastating disasters. Then there's the spread of infectious diseases and their carriers, or vectors as we call them, which include insects, mosquitoes, ticks, and more, spreading poleward as it gets warmer. And then there's the fact of increased flooding because the warmer air holds more water vapor. Increased flooding increases the risk of waterborne diseases, which predominantly affect children, especially under the age of five, in developing countries.

"And last but by no means least—in fact, I think this is the biggest one—climate change for many regions can be the final blow. You take a situation where there is devastating poverty, hunger, disease, lack of political stability, corruption, civil conflict, and strife, and you add an event that climate change has made worse or more frequent. And when a state fails, when a refugee crisis occurs, what's the first thing to go? Basic health. All of a sudden having a baby becomes a life-threatening situation. Running a fever could mean your death. So, to care about climate change, we don't have to be a certain type of person, we don't have to be a climate scientist, we don't have to be a medical professional, we don't have to be an activist. Whoever we are is already the perfect person to care about a changing climate. And the last, and most important, thing is this: There are solutions.

"This is a system-wide issue. If each of us individually did what we could in our personal lives, that in itself would still be not enough to fix the problem. That's why when people ask what's the most important thing they can do, my answer is this: Number **one**, talk about it. When you ask every county in the whole country, as we have done, 'Do you ever talk about this?' almost every county, except San Francisco, says, 'No.' If you don't talk about it, why would you care? And if you don't care, why would you ever act? And so that's why what we're doing here is so important.

"Number **two**, join organizations that share your values and amplify your voice.

"And number **three** is vote, because if we have to change the system, we have to vote."

As Ai-jen spoke, I thought about how the Green New Deal calls for elevating the value and dignity (and salaries) of low-carbon jobs like caregiving, nursing, and teaching.

The Reverend William Barber II was our next guest. I have held the reverend in awe ever since 2014 when I started hearing about his Moral Mondays civil rights protests in North Carolina's state capital. Like a lot of people, I was curious about what kind of man was able to bring together people across class, age, and race to fight injustice in the name of morality. He was building what the historian Timothy Tyson has called "a statewide interracial fusion political coalition that has not been seriously attempted since 1900." And then, on top of that, in 2017 he began the Poor People's Campaign: a national call for a moral revival in honor of the original 1968 campaign founded by Martin Luther King Jr.

"I was thinking just a moment ago about Dr. King," Reverend Barber said. "Dr. King said, 'If we don't learn how to live together as brothers and sisters, we certainly will perish together as fools.' Well, if we don't learn how to stand together and fight against what's happening with the continuing destruction of the climate through fossil fuel burning, then we may burn as fools together. And that's very, very real. I come from a community that suffered a five-hundred-year flood in eastern North Carolina. I'm just fresh from the Apache lands of Oak Flat in Arizona, where a company wants to drill down and take copper out of the land knowing that only 2 percent of what they bring up will be worth anything and the other 98 percent will poison the aquifers. I'm just leaving Cancer Alley, where we met a lady who said, 'I want you to know as you travel around the country, tell the people that I had five members [of my family] die in four weeks from the effects and the poisons of plants all along Cancer Alley.'"

When the time came for questions from our livestream audience, Stacy from Texas told us that all of her children have asthma and it seemed related to the environmental impact of the oil industry. She wanted to know if human illnesses are caused by fossil fuel extraction. Dr. Steingraber took on that one.

"Public health problems associated with drilling and fracking definitely include asthma. They also include poor birth outcomes. We're talking about babies that are born too early. That's the highest risk for infant mortality and for disability. Living near a place where oil and gas is extracted, if you are a pregnant woman, will put your child at risk. If we look at 1.1 million infants born in Pennsylvania, we find indications of poor health and significantly lower birth weights of babies that are born to mothers who live near fracking sites. An emerging study has found elevated levels of barium and strontium there. We know that those heavy metals found in the bedrock come up when oil and gas is fracked out of the ground. We see these in urine and hair samples of indigenous women living in areas of intense fracking activity in northwest

British Columbia. We know that those can harm babies. In other words, we're exposing women to substances that we know can sabotage human pregnancy.

"And as we went to press with our study, we are following twenty-seven cases of a rare childhood cancer called Ewing's sarcoma. It's a kind of a rare bone cancer that tends to strike young people in four counties in southwestern Pennsylvania. Six cases happened in the same school district, and this is a cancer where typically we would see only 250 cases all the way across the country. And this is an intensely fracked area of Pennsylvania. We didn't see these cancers before fracking. Now we do. We are literally killing our children."

A question came in from Washington State asking how climate change impacts mental health.

Dr. Hayhoe explained, "We're starting to realize that when we think about what's happening to our planet, we realize that in fact it's not the planet that's at risk; the planet will still be orbiting the sun. It is us that are at risk. That is a profoundly scary thought, and our reaction to that is fear, and we are seeing this fear, and we are seeing this anxiety, increasingly among us. Psychologists are looking at what they now call eco-anxiety.

"They're looking at our sense of grief and loss that the places we grew up in are changing in front of our eyes. In Alaska, I was with a woman who's lived in Juneau for decades, and she said, 'Those mountains always had snow on them, and now just this year they

LEARNING TO RESPECT OUR UNDERGROUND ECOSYSTEM

Dr. Sandra Steingraber

"We need a kind of cultural shift in the way we think about coal, oil, and gas. What if we didn't think about them as fuels? What if we thought about them in some other way? The mineral underworld of our planet—a symbol of everything devoid, dead, and inert—is not just fire and brimstone. The magma-warmed rocks deep below the sun-lit surface of our planet are actually an ecosystem. There's a living ecosystem under the ground that serves as a habitat for deep-life organisms, many of whom occupy an ancient domain of life called Archaea that can form complex colonies and alter their habitats to suit themselves—just as living organisms do here on top of the earth's crust.

"The carbon-rich geological strata that contain oil and natural gas are especially rich with a diversity of metabolic, reproducing life-forms. This fact is a problem for drilling operations because these organisms can flourish inside the pipes and well casings and interfere with the flow of oil and gas. This is why the extraction technique called fracking—which uses water as a club to shatter the bedrock and extract the gas and oil bubbles trapped inside—relies on powerful cellular poisons called biocides to exterminate deep-life organisms. In essence, fracking requires an underground pesticide-spraying program to bring the oil and gas to the surface. That's one reason why fracking fluid is so toxic and a threat to our drinking water aquifers.

"What if we thought about the rocks below our feet as part of creation in need of protection? What if we saw oil and gas extraction as a process of de-creating and desecrating the entire planet—swinging a wrecking ball at both our climate system and our deep-life ecosystem? Understanding that our earth's biosphere extends deep into the dark heart of our planet gives deeper meaning to the necessary task of keeping so-called fossil fuels in the ground."

don't, and I'm missing a part of what makes this place my home.' So, yes, we are being affected, and here's the thing: Fear can paralyze us. We need hope. We need a vision of a better future. We need to interact with each other because talking to each other, taking action ourselves, that is what gives us the hope that we need to fix these things."

I jumped in: "Over my many decades of activism, what I have found repeatedly is when my body is feeling despair, the moment I step into community and take action, the despair lifts. I was very depressed before I moved to D.C. and worked with Greenpeace on Fire Drill Fridays, and the minute I took action and did everything I could with my entire body, together with wonderfully smart experts and poetic experts and others, the despair lifted. Activism and community. The community that activism brings us into can be the antidote to despair."

The Rally

The next day, the extra-large Lutheran church basement, where we gathered before the rally, was crammed with red. Suddenly it seemed that every woman had a red coat and we were all on fire. So many friends were there, as well as men and women for whom this was their third or fourth Fire Drill Friday who had become friends. Annie Leibovitz was back, camera in hand. Roshi Joan Halifax, abbot of the Upaya Zen Center in Santa Fe, where I have experienced the Rohatsu Sesshin, the intense, transformative, eight-day silent, formal Buddhist meditation; Dolores Huerta and her daughter Camila, who, as a child, had been a camper at my Laurel Springs children's camp; Gloria Steinem, Eve Ensler, Pat Mitchell, Jodie Evans; Vassar Seydel was there with her mother, Laura Turner Seydel, and sister Laura Elizabeth.

I was astonished to see Lois Gibbs there. In the 1970s, Tom Hayden and I had gone to Love Canal, a community near Niagara Falls, New York, built on top of a toxic chemical dump by a subsidiary of Occidental Petroleum. Lois had been a shy housewife who organized her neighborhood to battle local, state, and federal governments, eventually forcing them to evacuate more than eight hundred families and clean up the toxic dump. For me, Lois had always been a prime example of women's capacity to transform into warriors when it comes to the health and safety of their children and neighbors. I was very moved to see her.

There were fifty-five nurses with the Alliance of Nurses for Healthy Environments, all wearing their white nursing coats and red berets. The place was pulsing with loving communion.

As we marched to the rally, I thought how different it was from our first uncertain, ragtag time back on October 11. We were larger now. We knew the chants, and our voices were stronger. We knew we were having an impact. People had come from all over because this is where they wanted to be.

Firas Nasr interviews Jane for the Fire Drill Friday livestream they are recording.

I opened the rally telling people that the next day, winter solstice, is celebrated by people around the world as the beginning of the return of the sun. Darkness turning into light.

"The solstice reminds us that even in the darkest time, the sun is not vanquished. Out of darkness, light is born. Out of disappointment and despair come new courage and new hope. At winter solstice, a metaphorical seed is planted in us. Over the coming year, this seed, planted in the darkness of winter, will sprout and grow, becoming something manifest in our lives at harvest time next November. Let's think about what we would like to bring forth to birth this coming year.

"I'll tell you what I want to birth, with your help, and with the young climate strikers and all environmentalists together, let's birth an army. An army of people ready to commit themselves to activism on behalf of the planet. The planet needs an army. Our children and grandchildren need an army. All the precious species teetering on the brink need an army of us ready to take the next step, to leave our comfort zones, as Greta has called us to do.

"But maybe, by doing this, we will be in a more profound place than comfort because we will know we are putting our bodies in alignment with our deepest values. And we will know that we are not alone but with an army of like-minded brave and determined people. We will grow in the next year to a mighty army because we have to. Now you may say that you came to celebrate my birthday with me in this wonderful activist context, which has gained so much traction, more than I ever imagined—and I appreciate this. But I think that we are all here today because we know what lies ahead and we need to be together now as this year draws to a close and a new, critical year dawns. We need to be together for collective fortification, for sustenance, and for love. And I'm very grateful. So let's stay together and love each other. And let's make the circle wider and the tent bigger."

Our first speaker, Heather McTeer Toney, was the first African American, the first female, and the youngest person to serve as mayor of Greenville,

Heather Toney speaks to the crowd.

Mississippi. In 2014, she was appointed by President Barack Obama as a regional administrator for the EPA. She is now national field director of Moms Clean Air Force.

Heather personalized her climate message. "I have a question for all of you. Have you ever heard your mother say, 'We need to talk'? At that moment, the rest of your life, the trajectory of your very existence, hangs in the balance. Well, take that feeling and multiply it by 1.2 million, because that's the number of moms with Moms Clean Air Force. And today we have a message to deliver. We're saying right now, Congress, your mothers are telling you, we need to talk.

"We need to have a serious conversation about the utter failure of your agencies to protect the health of our children. You've shown us you're willing to reverse and roll back regulations that directly impact their well-being, and that's not acceptable. So we need to talk. We need to talk. I'm from Mississippi, so the way we say it is, we need to have a come-to-Jesus meeting."

Laura Turner Seydel, who fifteen years ago helped found Mothers & Others for Clean Air in Atlanta, also focused on children in her remarks. She cited a

UNICEF report that estimated that children bear 88 percent of the burden of disease from climate change. In the city of Atlanta alone, Laura's organization estimated, the costs to families where the children suffer from asthma, with kids missing school and parents missing work, added up to $50 million a year.

Laura Seydel speaks to the crowd.

"What is the price of our children? Right now, our government is in the back pocket of polluters who are making our children and our communities sick. The only thing standing in our way are the politicians who refuse to acknowledge the basic science of climate change. What can we do? As my hometown hero Congressman John Lewis advises, we can create trouble, good trouble, necessary trouble. We can and we must. When it is all said and done, don't we want to be able to tell our children, our grandchildren, our nieces, our adopted children, that we did everything in our power to create a more just and equitable and livable planet?"

Everyone laughed and cheered when Gloria Steinem, a woman who has transformed our culture and so many millions of women's lives, said, "I am here to talk about only one of the women this orange man is

Gloria Steinem speaks.

disrespecting, assaulting, and endangering, a woman none of us can live without, and her name is Mother Nature." She described all the rollbacks, cancellations, ban liftings, revisions, and reversals driving nails further into the coffin of Mother Nature.

Gloria then movingly introduced Dolores Huerta, lifelong organizer of farmworkers, from the San Joaquin valley of California.

"South of Bakersfield the air was so contaminated that the children could not go out for recess. The parents went door to door. They passed a bond issue to build a brand-new gymnasium for their middle school right there in Weedpatch, California. These parents are farmworkers. They got together to protect their children. So that's what we have to do. We have to go out there and organize and we have to wake up the politicians to protect all of the children on our planet and stop global warming."

WHO WILL MAKE THE CHANGE?

Roshi Joan Halifax

"We know, firsthand, the power of the people when we organize. None of us, nor can our children, nor can our grandchildren, escape the effects of the climate catastrophe. Let us together join with all who are affected deeply and move the dial toward peace.

"Fossil fuels are a dangerous, finite, dirty, and destructive source of energy. Friends, we live in an interdependent world, and we cannot deny how damaging this energy source is to the individual and to the collective health of all species. Extreme climate heat is also linked with aggression and connected with violent conflict and forced migration, another source of profound trauma. And then there is the pernicious suffering, on the psychological level, experienced by those who witnessed the terrible degradation of life associated with our climate catastrophe and the aggressive assaults on the dignity and well-being of those who raised their voices.

"We must ask, 'Who will make the change?' And clearly, every one of us. No matter our skin color, no matter our age, we must all rise up. We face either the very real possibility of a planet on hospice, driven by an energy system that is the epitome of capitalism with extreme exploitation and racism at its core, or a profound opportunity to shift our economic system that we haven't seen since the abolition of slavery.

"So you and I must take action now. If there is to be a viable, morally grounded, and healthy future, we have to hold accountable those who are stealing the future from our children and our grandchildren. We must rise up together in solidarity with young people, indigenous people, people of color, and meet this collective crisis with committed, compassionate action. We must not cower behind walls of privilege that we erect out of fear. Fear, denial, and futility are no options now. Rather, we must vote. We must act. We must wake up, and we must wake others up and be a revolutionary force for a healthy future for all beings."

Dolores Huerta speaks, as Rosanna Arquette looks on.

And then Dolores shouted, "Who's got the power?" We responded, "We've got the power!"

Eighty-nine-year-old Dolores, with her more than sixty years of organizing, knew how to rally a crowd. And on the heels of her energy, the world-renowned a cappella group Sweet Honey in the Rock mounted the stage to sing me "Happy Birthday."

Then Reverend Barber moved slowly up a ramp to the stage, leaning heavily on his cane, but once in front of the mic the man seemed to grow taller, his back straightened, and his voice came from the core of his being as he knit together all the issues—poverty, racism, violence against women, the rape of the earth, climate change, homophobia—yet he ended on a note of inspiration, common purpose, and a call to act.

"And if we organize together, if we fight together, if we stand together, we hold the power to change the political calculus in America," Reverend Barber said. "So together we must stand for love and truth and

The Reverend William Barber cheers with Jane, alongside Roshi Joan Halifax.

justice and for the saving of creation. We can't stand down now and not ever. What this is now that we see, it doesn't have to be. We can turn it around."

After the rally, as we marched to the Hart Senate Office Building, I saw Gloria Steinem and Dolores Huerta on my left, Ai-jen Poo, Reverend Barber, and

Dolores Huerta, Gloria Steinem, Roshi Joan Halifax, Jane, Ai-jen Poo, Casey Wilson, and Laura Seydel march to the Hart Senate Office Building to engage in civil disobedience.

Heather McTeer Toney on my right, and it felt like a turning point. All these iconic movement leaders coming together on behalf of the climate, of our future.

As we sat on the marble floor of the rotunda, I looked up at Alexander Calder's black sculpture **Mountains and Clouds** towering over us. I knew Sandy when I lived in France in the 1960s. He lived on a farm where he gave lodging and support to American servicemen who were war resisters, the very men who had opened my eyes to the realities of the Vietnam War. I knew Sandy would be proud of what we were doing now.

The entire group pictured occupied the rotunda and risked arrest.

Gloria Steinem is arrested.

Roshi Joan Halifax is arrested.

There were a lot of us there that afternoon, hundreds, and the police took their time arresting and processing us. Suddenly Reverend Barber began to sing "We Shall Overcome" in a rich, deep voice, and everyone was so happy to join in.

Some say we were held in the Capitol Police detention warehouse for twelve hours. Maybe it was that long, but it didn't feel like it. There were approximately 143 of us that day, with women outnumbering men by four to one.

As before, men and women were separated by an aisle, and we were supposed to remain in our seats and not mingle or make noise. But after a few hours, we began to move around, and the police didn't stop

Catherine Keener and Rosanna Arquette wait to board the police vans after they are arrested.

Annie Leonard and her daughter, Dewi.

us. For the most part, the men were subdued, whereas the women were abuzz, talking, organizing, making connections.

I got to know Aileen Getty, who was part of the newly created Climate Emergency Fund, set up to support the young climate strikers and other big climate actions. I had known Aileen's father, Paul Getty, while filming **Barbarella** in Rome in 1967, and she and I made plans to get together in the future.

Sandra Steingraber's college-age son, Elijah, crossed the aisle to tell me that due to prolonged drought in the world's tropical forests, the wood used to make classical instruments such as violins and woodwinds lacks its former moisture levels. As a result, modern instruments no longer emit the same quality of sound.

During the long detention, Karen Nussbaum, curious about the many people there whom neither of us knew, spoke with a few of them and later relayed to me what she'd gleaned. Cheryl Barnds is a mother of three from Takoma Park, Maryland, who had never been politically active before she started coming to Fire Drill Friday rallies. She now volunteers at pre-rally meetings and is recruiting people to continue Fire Drill Fridays in Takoma Park.

Kaysha, nineteen, and her aunt, Shaila, both of Middle Eastern descent, came to D.C. on a bus from Michigan, organized by Kaysha's friend, Sohil, also nineteen. They told me Sohil follows me online from school in Michigan, which is how they knew about Fire Drill Fridays. When they decided to get arrested, Shaila insisted they get an okay from Kaysha's mom.

Then there was Sandra, who turned fifty-eight on the November 1 Fire Drill Friday devoted to women and climate. Sandra always celebrates her birthdays with something special, like a sixty-mile bike ride. She made Fire Drill Fridays the birthday event for this year, and she had come back for mine.

As you may remember from chapter 2, the team decided back in September that Fire Drill Fridays were targeting people who were concerned about the climate crisis but had not yet been moved to activism that included civil disobedience. Hearing from Karen

Karen Nussbaum gives a playful kick toward the crowd as she gets arrested.

about the Cheryls, the Kayshas, and the Sohils gave us proof that we were succeeding.

At one point, a young black policewoman approached Gloria Steinem, shyly pulling out of her pocket a copy of the U.S. Constitution (they are all required to carry one), and whispered to her, "I may lose my job over this, but will you sign my Constitution?" I wish I'd had my phone to take a picture of Gloria, her hands bound with plastic handcuffs, signing the little Constitution.

At about 10:30 p.m., amid cheers and applause for their willingness to go to jail, I watched Karen Nussbaum and my new Fire Drill Friday friend Erika Berg get patted down for a second time and taken away. Clearly a decision had been made to not lock me up along with them, although I was as eligible as they were. I was the very last to be released, and I would have been more disappointed not to have turned eighty-two behind bars had detention not turned into my best birthday party ever.

What Can I Do?

Start by taking care of your own health and encouraging those close to you to do the same. We all need to be healthy for our collective struggle to make change and have the energy required for movement building and civil disobedience! A changing climate means you'll be facing stronger, longer heat waves and greater exposure to what's called vector-borne disease (like malaria from mosquitoes) as these heat-loving creatures find more habitat in which to breed and spread. Wear mineral sunscreen that is safe for ocean life, drink plenty of water, and wash your hands frequently with regular soap; unless really needed, best to avoid those antibiotic cleansers that kill the good bacteria, too.

Make a special effort to get to know your neighbors, if you don't already. Elderly folks and other immune-compromised individuals are at the greatest risk when things like heat waves or other extreme weather events occur. Check in on your high-risk neighbors frequently; make sure they have the food and water and medicines they need, or an emergency plan in place in case they need medical attention or to evacuate.

Community is one of the main things that is going to see us through the upcoming climate-related disruption. As Reverend Barber reminded us, "In community, we don't ignore the pain. We address it together."

For more community-based, collective actions, you can join organizations that "share your values and amplify your voice," as Dr. Katharine Hayhoe put it. We will need much more caregiving and nursing as the climate and its health-related effects, including pandemics, intensify. Health-care workers are not paid anywhere near enough to take care of themselves, never mind their patients. You can support National Domestic Workers Alliance and make sure the Green New Deal and health justice are priorities in the climate groups you're already involved with. Here are some great examples; choose one or more of these, research others, or start your own local effort!

If you're a nurse, contact Alliance of Nurses for Healthy Environments, which has been involved with Fire Drill Fridays since its launch in Washington, D.C., and learn about its Climate Change Committee.

If you're a health-care provider or work with any health-care institution, join Health Care Without Harm's Climate Challenge to commit to climate-smart health care. Health Care Without Harm has lots of resources to ensure that hospitals are at the forefront of climate solutions like committing to renewable energy and non-incineration waste management and supporting local healthy food.

You don't have to be a health-care provider yourself to check out Health Care Without Harm's Climate Action Playbook for guidance on making sure your local hospital is a climate leader.

Holding government officials accountable is another crucial action you can take. Workers are paying the cost of climate change with their health. We need to stand together to protect those on the front line; ask your congressional representatives to support the Workplace Heat Protection Bill, which you can learn about from the advocacy group Public Citizen.

Lily Tomlin and Rolando Navarro chant with Jane
on the Capitol steps after the rally ends.

CHAPTER FOURTEEN

Forests and Climate Change

It was Christmas week, and Washington was pretty much shut down. None of us were sure what kind of turnout we'd have for the Fire Drill Friday coming two days after Christmas. But this was the Friday that Lily Tomlin was going to be part of the action, and I knew that she would be a draw. The **Grace and Frankie** Christmas Arrest Special! I'd been talking about it for weeks, and I was so looking forward to being with her again. I love Lily and for the last six years I've been able to spend half the year with her filming our show. After we film this final season, I will definitely have withdrawal symptoms. Her coming to D.C. wasn't just a boost for Fire Drill Fridays; it was my Lily fix after a long time without her.

Just as we had focused on food for the Thanksgiving holiday, it seemed fitting to have our Christmas focus be forests. You know, Christmas trees.

Forests have always been my happy place, and I've

hiked or ridden horses through many: the giant sequoias and redwood forests of Northern California, the bamboo forests in Japan, the rain forests in northwestern Australia, and the temperate rain forests in Russia. Each forest has a different personality, different stories to tell, different flora and fauna living under its canopy.

People sometimes call forests "the lungs of the planet." Half the oxygen we breathe comes from the forests, with the other part coming from the sea. Forests do more than that by storing carbon in their trunks and leaves and roots. Forests create their own weather. The moisture that evaporates from their leaves forms clouds, makes rain, contributing to the complex balance of nature.

When we mow down forests, the systems that support life become unbalanced. This is very true in the Amazon, where a steep increase in fires made global headlines in recent years. As the Brazilian government has allowed, and even encouraged, burning the Amazon rain forest, more extreme droughts linked to deforestation have threatened water supplies for millions in Brazil, including those in faraway cities like São Paulo.

We've already lost 30 percent of the planet's forests; 20 percent of forests are degraded, and much of the rest is fragmented, leaving only 15 percent of the original forests intact. That is a scary figure. The focus of the teach-in was to learn what we can do to stop degrading and destroying this important resource.

The Teach-In

As the Thursday teach-in began, I told our viewers that one of the best Christmas presents I got was finding out how many people were actually watching the livestreams—more than 100,000—and that they were letting us know that they were learning a lot. "I can't tell you how happy that makes me," I said, "because we're just sitting here looking at a tiny little iPhone, about ten feet away, and it's kind of weird to talk at it. And so now that we know that there are actually people watching us and finding it interesting makes it all thrilling for me."

I was grateful to our guests, Rolf Skar, senior campaign strategist for Greenpeace, and Hana Heineken, from Rainforest Action Network, for taking time away from family at the holiday. As Greenpeace's former forest director, Rolf has decades of experience leading campaigns to end deforestation in Indonesia, the Brazilian Amazon, the Congo basin, and the Canadian boreal forests. Hana attacks the problem from the business end, holding banks and investors accountable for backing companies that destroy the rain forests and the climate and that violate human rights.

Hana began by describing the devastation to forests just in 2019. "California was on fire this year, and it's not just California," she said. "In the Amazon, about 4.6 million acres of rain forest are on fire. Indonesia as well, over 2 million acres of forest were on fire.

Hana Heineken speaks.

The Congo, considered the second-largest tract of rain forest, was also on fire. In Australia the bushfires actually have been burning 7.4 million acres. And then the Arctic. There are fires in the Arctic as well. Clearly these fires are being caused by climate change, global warming, but they're also deliberately set by those who want to burn the forests to clear them for agriculture and cattle ranching."

Indonesia's arsonists want to clear the forests in order to grow palm oil trees, because palm oil has become a valuable commodity. But when they burn the forests, they destroy the peat lands, dense wetlands with layers of moss and shrubs that store tremendous amounts of carbon. Hana noted that when most people think of climate change, they focus on reducing fossil fuel consumption, but fossil fuels aren't the only source of carbon emissions.

"So this is a highly neglected carbon emission, and in fact the carbon emissions that resulted from this year's fires in Indonesia alone made Indonesia the sixth-largest emitter of carbon in the world, actually much greater than the carbon that was emitted in the Amazon," Hana said. "We need to protect the carbon that is stored in these forests, in these peat lands, and allow them to continue to absorb all of this carbon."

Rolf added that deforestation and degradation of forest cause more damage to the environment than all the cars, trucks, trains, planes, and ships in the world combined. "It's an enormous amount of carbon we're talking about. Not just in the trunks and the roots and the leaves, but, as Hana started to talk about, peat lands and soil. In some of the biggest and most important forests on the planet, the carbon's actually in the soil. That's the case in the boreal forest, which stretches across Alaska and Canada. In Russia, it's like a green halo on top of the planet. When you disturb those soils, you're tapping into this carbon piggy bank that's been building up for hundreds or even thousands of years."

I asked Rolf to explain what is meant by "fragmenting" in relation to forests.

"Fragmenting is when you allow roads through the forest or permit mineral extraction there. When you start fragmenting the forest, the edges dry out," he explained. "Some of the species don't exist anymore; sometimes those species are important for the good of the whole. Those things really add up, and we're learning there's a big climate cost associated with them."

"Does replanting mitigate the problems caused by deforestation?" I asked.

"I think it's good to do," answered Hana. "But along with planting trees, we need to save the trees that are currently standing, especially in those vulnerable areas like the tropics, but really everywhere. Let's plant trees. But let's also stop the deforestation."

"Hana's right. We've got to stop the bleeding first," Rolf said. "Which means ending deforestation. Some people have said we'll never achieve that. What they don't realize is that it's inevitable. It's just a question of whether we wake up one day and it's all been wasted and it's too late, or whether we work together now to develop solutions for people on the planet. We need to tackle the fundamental drivers of deforestation. At the top of that list: industrial agriculture, meat and dairy production, and pulp and paper plantations. Palm oil plantations, even rubber, are emerging as a fundamental driver of deforestation."

"I'm sure many of you have seen 'palm oil' written on products like soaps, snacks, shampoos, and whatnot," Hana said. "If you look at paper products like toilet paper or [tissues], if it's coming from Indonesia, it's probably related to the destruction of these peat lands. The consumer has a real opportunity to influence and challenge that supply chain through boycotts and divestment.

"In the 2000s, we targeted Citigroup. We were able to get them to adopt one of the first policies on environmental and social issues that led to a whole chain of other financial institutions adopting policies as well. And both Greenpeace and Rainforest Action Network have also been working with Unilever and Nestlé. These companies have committed to what we call 'No Deforestation, No Peatland, No Exploitation' policies, which means they will commit that their supply chains are free of deforestation. In order to achieve that commitment, they need to have a full understanding of what their supply chain looks like. Which trader is it coming from? Which producer is that trader buying from? Where is it actually being grown? Some of these companies, like Unilever, have made a huge effort to clean up their supply chain."

Hana said 2020 was an important year for forests, a year when four hundred companies promised to eliminate deforestation from their supply chains. Also, by 2020 the United States, Canada, and the EU promised to halve forest loss and end deforestation completely by 2030. "We're going to do our best to make sure those companies are sticking by their commitments and also helping to achieve this global goal that all these countries have signed up for," Hana said.

"That's why we need people power now more than ever. The only way that this works is if we mobilize your energy, people's energy, to hold both the companies and our governments accountable," Rolf said, looking straight at the audience represented by the little iPhone. "There are all these pledges on the books about 2020, or even 2030, yet companies are not making enough progress and governments are falling short on the pledges they've put on paper. We don't need pledges on paper. We need results on the ground. People power has gotten these companies' attention. We've moved this idea of ending deforestation from the abstract into concrete policies. We have to let them know that the whole world is watching and that we're really determined about the future of this planet right now.

"The Amazon rain forest is the largest in the world. And long story short, if you just protect and uphold and respect indigenous peoples' rights, you've got a good chance of protecting forests because these are folks who've lived in these lands since time immemorial; they depend on the forests for their way of life and very existence, so they tend to manage them sustainably. Research show that deforestation rates inside forests legally managed by indigenous peoples are two to three times lower than in other forests. The president of Brazil, who just got into office in January 2019, ran on a platform of tearing apart protections for indigenous peoples and traditional communities, of doing whatever he needed to pay back big industrial, agriculture interests who helped get him elected. We can send a message to the Brazilian president that these kinds of practices aren't good for business. If you want to sell your products to the U.S., the EU, other countries, we're not going to want them if they are stained with the blood of indigenous communities and stained with climate change and forest destruction."

Rolf noted that forests provided many benefits beyond mitigating the effects of climate change, some we were still discovering. I agreed.

"We haven't even identified a lot of the species, or the medicines, that could come from there," I added.

"Even in the Pacific Northwest of the U.S., there was a tree that was thought of as a trash tree," Rolf said. "It grew in an old-growth forest. It's called the Pacific yew. And the bark, it turns out, was instrumental in developing a drug that helps treat breast cancer. Many chemicals found in the plants and parts of trees that grow in the Amazon have been used to develop medicines. We literally don't know what we're ruining and what sort of charges we're putting on our ecological credit card for future generations to pay."

The time came for questions, and the first questioner, from Colorado, wanted to know what U.S. policies are protecting forests.

Rolf answered that when a forest is on private land, what happens there is largely up to the landowner, but the biggest wild forests are on public land and belong to all Americans. "That's where you find the big wild forests, the big wild rivers, and most of the wildlife. And on those lands, there's a complicated mix of policies that protect those forests. For example, take the national forest in Alaska, one of the biggest and most pristine and environmentally critical temperate rain forests on the planet, home to many rare and endangered flora and fauna. They are trying to road, log, and mine in that last big wild forest. And it's important that we push back because those are our lands."

"You're talking about the Tongass National Forest!"

I exclaimed. "Several Fire Drills ago, a delegation of indigenous women in their traditional regalia from the Women's Earth and Climate Action Network joined us. We hadn't expected them, but I invited them to speak. The women had traveled to Washington from a town of eight hundred people in southeast Alaska to advocate on behalf of the Tongass. They were fighting to stop the Trump administration from logging. I'll never forget them. They closed the rally with their keening songs and drumming, pleading for the forest's future."

A viewer in California wanted to know how the food we eat is related to deforestation.

"I got into the forest-saving business thinking I'd be talking about chain saws and bulldozers," Rolf said. "And it turns out for years I've been talking about agriculture and food. In the Amazon, they're getting rid of the forest and replacing it with a monoculture of soy plantations. Not to make tofu, but to feed the animals. It's a very inefficient way to feed people. Cattle ranching takes up a lot of space. I'd say, in general, we need to rethink the industrial model of agriculture."

"What are the big fast-food chains that need public pressure the most? Is McDonald's the number one villain here?" I asked.

"McDonald's helped create, along with Greenpeace and a bunch of other partners, what's called, uninspiringly, the Soy Moratorium," Rolf answered. "The idea was don't buy soy from places that have recently deforested, and that actually has held back a lot of deforestation in the Brazilian rain forests. Yum! Brands is the parent company to KFC and to Taco Bell. They've done some things, but it's more talk than walk with them. Burger King has been ignoring these problems. Now is the time to speak up." Rolf was again looking straight into the camera.

"The stats show that about 70 percent of tropical deforestation is actually being caused by these companies," said Hana. "Most of these companies are publicly listed, and they are beholden to their shareholders. And if the shareholders start saying, 'You're violating human rights' or 'You're deforesting the tropical forests,' that can have a huge impact. Our helping to mobilize shareholders to pressure these companies has had a huge impact on company behavior."

Torres, from Iowa, asked, "Yesterday was Christmas and my family buys a real tree every year. Should we stop doing that?" I was glad for the question because I'd been wondering the same thing myself.

Rolf said, "Christmas tree farms aren't one of the leading causes of deforestation. If folks want to have a real Christmas tree in the house, I think that's fine. One thing to look at with Christmas tree farms is whether or not they're using lots of pesticides. Some of them do that, and the toxins can end up in drinking watersheds and the rest of it. But oftentimes, they're already located on areas that were farms or pasture."

A questioner from Vermont asked how deforestation relates to animal habitat, like with tigers and palm oil.

"Destroying tropical forests has a huge impact on

biodiversity," Hana responded. "Fragmenting those forests is especially harmful for animals like tigers that need a large area in which to roam. So in Indonesia, for example, the orangutans are declining rapidly because they are losing their habitat largely for palm oil. But it's not all doom and gloom. Rainforest Action Network, with some local civil society organizations and other stakeholders, is in the process of preserving an area in Indonesia called the Leuser Ecosystem, which is called the orangutan capital of the world. It's the only place on earth where you can see the tigers and rhinoceroses and orangutans and the sun bear all living in the same habitat. An incredible place. Again, not foolproof. There's still a lot of loopholes and things we need to work through, but the deforestation in that area has been declining, which bodes well for the ability to have a real impact on the ground."

Laura from D.C. wanted to know what mining and minerals companies contribute to deforestation.

"Mining is a driver of deforestation, especially in certain regions," Rolf answered. "Gold in the Amazon—I saw it myself. It felt like going back in time and seeing California during the gold rush. It was the Wild West, with people showing up with guns, grabbing land, using really toxic methods of getting the gold out of the riverbeds and now those toxins, that mercury, getting into the rivers and the fish and people's bodies. There's a knock-on effect for mining: Sometimes it's the first thing that moves

THE EXTINCTION ADMINISTRATION

Rolf Skar

"Earlier we talked about the extinction crisis that humans have created around the world, and how we're losing things that we didn't even know were out there. Orangutans are one of our closest relatives. We share something like 97 percent of our DNA with them. I think they have the longest mother-child relationship among vertebrates in the world after us humans. They can stick together into almost their teens. They have to learn so much from their parents to survive in the forests. Think about what happens if the young are orphaned. We thought there were two species of orangutans, and then recently we found out there's a third. And it was immediately critically endangered. And to save these animals, we need to save their habitat. There's no future for wildlife without a home.

"We've nicknamed the Trump administration the Extinction Administration because it looks like that's what they're aiming for, maximum damage on our public lands, which largely have the best habitat remaining, especially in the lower forty-eight states. Attacks on the Endangered Species Act are a big deal. That might be our most important conservation law, certainly the most important for saving wildlife in the U.S. I mean, our national symbol, the bald eagle, came back from the brink of extinction because of that law. We've got many examples of the Endangered Species Act really working and working in a way that's good for people and the planet. They want to gut that. Not with a direct attack, but by taking away the enforcement of it. What good is a rule if it's never enforced? And that's what they're trying to do right now. They're trying to make sure that even though it's on the books, there won't be the teeth to it to enforce it, and we'll see how far they go before the next election."

into a frontier forest region like the Amazon. Sometimes it's the first thing that encroaches on indigenous peoples' traditional lands. And once those roads and those people have shown up from outside, it becomes easier for the illegal loggers to come in. And then the ranchers follow. So sometimes, as the saying goes, the first cut can be the deepest because it opens up this domino effect of deforestation."

"I'll add in Indonesia, where we've also found strong links between coal mining and deforestation," said Hana. "And we actually traced wood coming from coal mines to the Tokyo 2020 Olympics and their construction of new venues. So that's a whole other story. But it's a real issue."

The Rally

My assistant, Debi, was back from California that evening, and the next morning she and I picked up Lily at her hotel at 9:00 for the pre-rally gathering. More people than I expected had signed up to engage in civil disobedience. As we marched to the rally, we chanted,

> **People going to rise like the water,**
> **Going to shut this crisis down.**
> **I hear the voice of my great-granddaughter**
> **Saying, "Keep it in the ground."**

Hana Heineken, Rolf Skar, Gaurav Madan, Jane, Lily Tomlin, and Rolando Navarro walk to the rally site in the morning.

I introduced Rolando Navarro, with the Center for International Environmental Law (CIEL), who also had been president of the Agency for the Supervision of Forest Resources and Wildlife in Peru. For challenging the network of illegal timber mafias, Rolando and his family were targets of threats and retaliation. In 2016, Rolando had to seek exile in the United States, where he continues his work protecting Peru's Amazonian forest as a fellow at CIEL. He spoke to us through an interpreter.

"We know scientists have given the alert that if we stay on this path of forest destruction, we will be at a point of no return as a result of human activity and the impacts of climate change. The same alert has been given by environmental defenders in South America, from Peru, Colombia, Chile, Brazil, in Central America, from Guatemala and Nicaragua, but also from Africa, Asia, and Europe. There are women and men, human rights defenders, who have died fighting for this cause, who die every day because they have raised their voices in defense of our forests."

Then Rolf took the microphone and stirred the crowd with his call-and-response.

"The world watched in horror this year as the Amazon rain forest burned from human-caused fires set by land-grabbers emboldened by Brazil's new president, Bolsonaro."

The crowd booed at the mention of that name!

Rolf went on. "Burning a hole through the heart of biodiversity of life on our planet, trampling on the basic human rights of indigenous peoples, and calling into fundamental question the future of our biggest rain forest. This is . . ." He waited for the crowd.

And the crowd answered, "Not a drill!"

Rolf: "In Indonesia in recent years, the slashing and burning of carbon-rich peat rain forests have created plumes of toxic smoke, burning for weeks or even months, crossing international boundaries, visible from **Rolf Skar speaks.**

space. A recent study just showed that maybe 100,000 people, mostly children and elderly people with respiratory illnesses, may have died prematurely in one year alone in Indonesia, as a result of that deforestation. This is . . ."

The crowd: "Not a drill!"

Rolf: "When people, indigenous people and people in affected communities, dare to stand up for their lands and push back, dare to stand up for their fundamental human rights, to fight for their very way of life and existence, they're being targeted, persecuted, murdered. Since 2001, half as many land defenders have been killed as U.S. troops in Iraq and Afghanistan combined. This is . . ."

Crowd: "Not a drill!"

Rolf: "Climate change is extending fire seasons the world over, including where I live, in California. And Australia. There are the people who are left behind who are wondering how they are going to keep their children safe and how are they going to rebuild their lives. This is . . ."

Crowd: "Not a drill!"

Rolf: "Let's do it together. Let's say it one more time. This is . . ."

Crowd: "Not a drill!"

By way of introducing Lily Tomlin, I admitted that until a few years ago I hadn't known anything about the evils of palm oil. "But in the second season of **Grace and Frankie**," I said, "Grace, very innocently, was putting palm oil in the vaginal lubricant that she

Lily speaks; Hana Heineken is behind her.

and Frankie were going to sell. You know, the down-there challenges faced by older women. Well, when Frankie found out what Grace was doing, she went on a rant like Grace had never seen. She rubbed red paint all over herself and performed a sort of guerrilla-theater number to dramatize what was happening to the orangutans she so loved. It was so dramatic that I had to research it. Me, Jane, not me, Grace. And that's how I found out about the evils of palm oil. So it's only fitting that Lily is here with us today."

"I definitely won't be funny today," Lily said, and the crowd laughed. "I mean, I look around, and I know in my heart that I'm not the only tree hugger here. Some of the young people don't even know what

Lily Tomlin is led to the police vans after she is arrested and handcuffed.

that term means. It's just amazing, but sadly, hugging trees isn't enough today."

She talked about the devastating fires and the financial institutions that are wreaking such havoc on our world's forests.

"The largest investment company in the world, BlackRock, invests in the companies that do the deforestation, the drilling, and the fracking. It's beyond reprehensible. The heads of these companies, these supply chains, these traders, are CEOs. I think we should institute a new name for them, CEOgres."

Lily ended by saying that maybe we shouldn't buy real Christmas trees anymore.

"If you've got a real Christmas tree, save it till the little pine needles fall off and let it stand there as a symbol of consciousness. You'd love to be able to go and pull out that fake tree and put it right up, wouldn't you? With ornaments on it and everything? That's what we do at our house."

In real life, not just as Frankie and Grace, Lily and I like to rib each other. So when she sat down, I went to the mic and told people, "Well, Lily is just wrong there. There's nothing wrong with buying real Christmas trees."

Lily jumped back up and said, "Are you saying I'm wrong?"

"You bet you're wrong," I replied. "I learned at last night's teach-in that it's okay to buy real Christmas trees."

Lily tried to take back the mic to prove herself correct, and I wouldn't let her. She sat back down and turned her chair toward the corner of the stage like a chastised schoolgirl, while the crowd roared with laughter as I continued to explain why it was okay to buy a real tree.

DEFORESTATION EFFECTS

Gaurav Madan

Gaurav Madan is a senior campaigner
at Friends of the Earth and advocates
for corporations to stop financing
climate chaos through deforestation
and to recognize the land rights of
indigenous peoples.

"I want to talk a little bit about why we are here today. We know what we must do. We must phase out oil, gas, and coal and make fossil fuel industries a thing of the past. But we also must protect the world's forests. They go hand in hand. If we do not do one, we will not succeed. If we do not do both, we will not succeed in mitigating the worst effects of the climate crisis.

"Deforestation is not only fueling climate crisis. It is fueling unprecedented species extinction and gross human rights abuses around the world. The world's forests are at the epicenter of these intersecting and overlapping crises. Commodities like palm oil, pulp and paper, soy, and cattle are responsible

Gaurav Madan speaks.

for 80 percent of deforestation around the world. At the front lines of this are local, often indigenous communities that are facing a global epidemic of violence and murder. In 2018, three land and environmental defenders were murdered every single week. Imagine that. Three people killed every single week for defending what is theirs, for defending their way of life, their livelihoods, their family, and their future.

"This problem is not limited to South America. We see the destruction of the world's forests from Indonesia to Liberia to Brazil. So who is funding this? The same banks and investors that are financing the oil, coal, and gas industries.

"I want to talk about BlackRock, the

largest investor in oil pipelines, the largest investor in gas power plants, and the largest investor in coal mines. But if that wasn't bad enough, they're not only the largest investor in fossil fuels; they are also the largest investor in rain-forest destruction. They are literally fueling and financing the climate crisis. And they say they are a socially responsible investor. They've been called the conscience of Wall Street. But the reality is this is straight hypocrisy.

"They only support 10 percent of shareholder resolutions calling for climate action. When it comes to human rights, they voted last year for 0 percent of human rights resolutions. We have to hold the financial sector accountable, force them to protect our planet and do the right thing.

"I want to close on a personal note. I mean, we are all here in solidarity with our brothers and sisters who are resisting the takeover of their land. We are in solidarity with our sisters and brothers in the Midwest who are fighting the expansion of pipelines. This spring I will be welcoming my first child into this world. And I think that makes this a little more urgent for me. The future I want for her is one where everyone is treated with dignity and we have a safe and livable planet."

"I love it when Frankie's wrong and she hates it when I'm right and I am right in the case of Christmas trees, because Christmas tree farms, for the most part, are put onto already cleared and often degraded land."

With that, Lily and I hugged and marched with the crowd to the Capitol steps. I left the steps after the third warning from the police, but Lily stayed, having committed herself to risking arrest. I watched and cheered as my beloved pal got handcuffed and led away, a smile on her face.

What Can I Do?

We cannot survive without the forests—the lungs of the earth. To help the forests, you can start with individual changes, then broaden your impact by participating in social movements working for changes at the international level.

At the individual level, reduce your consumption of single-use products, particularly those derived from paper, wood, and rubber. When making food choices, avoid products that contain palm and soybean oil and avoid meat from livestock that grazed on deforested land or that were fed soybeans from deforested areas. If you can, eat a more plant-based diet of vegetables, fruits, and legumes. Your body will thank you, the forests will thank you, and the whole earth can B-R-E-A-T-H-E better. Of course, our individual food choices won't solve everything, but it's an important first step with many benefits.

Insist that retailers and restaurant owners make more forest-friendly choices available. Ask if the meat you're buying was raised sustainably, and be sure the forest-derived products you purchase are 100 percent postconsumer recycled content. When products are made from "virgin forest," insist that they are sourced in an environmentally and socially responsible manner. When you ask, be curious rather than insistent, because your server, or the clerk at the shop, may not have these answers or know how to get them. Your goal is to start the conversation, help spread the word, and remind the businesses that they depend on **you**, the consumer, and that you're keeping track. If they want you to shop there, they had better up their game.

In addition to speaking directly to retailers, you can speak directly to companies. It's easy to get a corporation's attention by tagging it in a post or tweet. You can also just give them a good old-fashioned phone call, or even write a letter, which might get more notice for the rarity and also because the time you took shows how serious you are. Demand that these companies commit to reducing deforestation through forest-friendly policies and that they follow through on those commitments.

These individual and market-based actions are not enough. Demand that your governments buy forest commodities that protect nature and respect human rights. Remind your elected officials that indigenous

peoples have been forest protectors since time immemorial and that their rights to their traditional lands
must be upheld, for their sake and for the planet's, as
recognized by the UN Declaration on the Rights of
Indigenous Peoples.

Call out the transnational banks like Union Bank
and big investment firms like BlackRock that fund
the destruction of nature that punctures the planet's
lungs. Make sure none of the financial institutions
and investment products you're involved with support
these extractive industries, and demand that your
government divest as well. You don't have to go it
alone. Groups like Greenpeace and Rainforest Action
Network have been fighting these battles for decades.
Support their work as well as the work of indigenous
organizations like the Indigenous Environmental
Network.

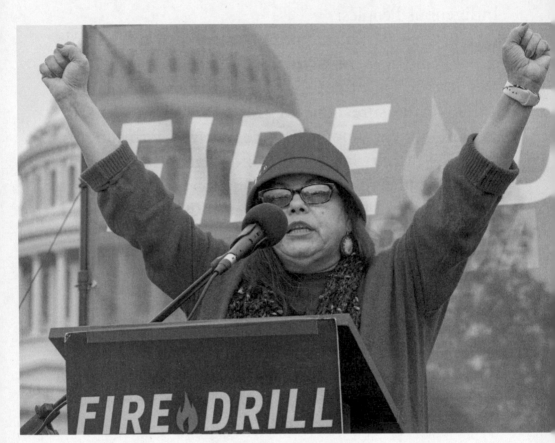

Donna Chavis speaks.

CHAPTER FIFTEEN

Holding the Fossil Fuel Industry Accountable

For twelve Fire Drill Fridays we focused largely on the effects of fossil fuels on our climate, on our environments, and on the communities at the epicenter of their toxic plundering. Finally, the time had come for us to address head-on the need to hold the fossil fuel companies accountable and the ways to do it.

Let me just start off by telling you something that I hadn't realized until I moved to D.C. for Fire Drill Fridays: I had not realized how many people—even self-proclaimed climate champions—refuse to embrace, let alone discuss, the need to reduce fossil fuels at the source.

Never in twenty-five years of international climate negotiations was there a meaningful discussion about stopping fossil fuel extraction. Even the Paris

Agreement and the Green New Deal resolution **don't mention the words "fossil fuels"**! Even though fossil fuels are the single largest driver of climate change, until very recently many environmentalists, including some of the big environmental organizations, seldom talked about fossil fuels and steered clear of any call to reduce them. They talked about a sustainable clean energy future. They talked about solar panels. They talked about windmills and conservation, but few talked about the need to confront fossil fuels. And when it comes to tackling climate change, U.S. politicians are no better. Very few have acknowledged the need to keep fossil fuels in the ground. I was absolutely floored when I learned this. Why would you be fighting the climate crisis and not talk about what's causing the climate crisis? It seemed so counterintuitive.

It took a while, but now I understand better: It's way easier and less controversial to avoid confronting the fossil fuel industry. I mean, think about it: Just about our entire economy depends on fossil fuels. Fossil fuels enabled our nation's development as an industrial powerhouse. The scientists are right to say that staving off a fossil-fuel-induced climate disaster "requires nothing less than rapid, transformational action." It's much easier to advocate building wind turbines than to set in motion transformational action to alter the economic sector that undergirds our society.

Also, fossil fuel corporations are very powerful economically, politically, and socially. Standing up to them requires a strong spine and a readiness to refuse their gifts. Fossil fuel corporations donate millions to political candidates. They also donate to environmental and other do-good organizations, sponsor community and cultural events, and even donate parks and baseball diamonds to the very communities they are polluting.

And that's why the shift to naming names has emanated from the most courageous part of the climate movement, starting with the frontline and indigenous groups who have no choice but to confront fossil fuel extraction and production because it is in the heart of their communities, where they live, work, play, and pray. And the call has been echoed by some national and international groups: 350.org and the fossil fuel divestment movement; Greenpeace and the networks of indigenous peoples' civil disobedience targeting fossil fuel infrastructure; the Sunrise Movement's demand that candidates sign the No Fossil Fuel Money Pledge and refuse contributions from the fossil fuel industry; Oil Change International's work to end public subsidies for fossil fuels; Greta and the young, courageous climate strikers. These are the forces that have been instrumental in making the shift in focus from uniquely on the demand side to the supply side: End fossil fuels. You can't solve a problem if you're not naming it. This is why Fire Drill Fridays' banners put getting off fossil fuels front and center and we never had speakers whose organization didn't support this demand.

Here's another mind-blowing thing I didn't fully understand until I got to D.C. and began studying in earnest: The urgency we now face would not have been necessary had the oil companies told the truth about what they knew fifty years ago.

Documents found by enterprising journalists and lawyers reveal that oil companies knew in the 1960s of the risk of severe environmental damage from global warming caused by their products.* A decade or two later, Exxon's and Shell's own scientists told them that the warming effects of their carbon emissions could double even earlier than previously predicted, causing ecological calamity, such as the disintegration of West Antarctic ice sheets, that would flood entire low-lying countries and farmlands, destroying entire ecosystems, and that new sources of freshwater would be required. They knew that global changes in air temperature would "drastically change the way people

---

*A 1968 Stanford Research Institute report commissioned by the American Petroleum Institute (API) entitled "Sources, Abundance, and Fate of Gaseous Atmospheric Pollutants" concluded, among other things, that fossil fuel combustion was by far the most likely "source for the additional CO_2 now being observed in the atmosphere" and warned that rising CO_2 would result in increases in temperature at the earth's surface that could lead to potentially serious environmental damage worldwide. Neela Banerjee et al., "CO_2's Role in Global Warming Has Been on the Oil Industry's Radar Since the 1960s," **InsideClimate News**, April 13, 2016; Benjamin Franta, "Early Oil Industry Knowledge of CO_2 and Global Warming," **Nature Climate Change** 8, December 2018, pp.1024–25.

live and work and that the changes may be the greatest in recorded history." Their reports showed that the American Midwest and other parts of the world could become desertlike.

And you know what Exxon's officials said? "This problem is not as significant to mankind as a nuclear holocaust or world famine,"* they asserted while continuing to drill.

Exxon, Shell, Mobil, and others knew that their products wouldn't stay profitable once the world understood the risks, so they used the same consultants that the tobacco companies used to launch a huge communications effort to develop strategies on how to fool us. The difference is that the tobacco companies were primarily harming smokers. The fossil fuel companies are harming the entire planet and all its inhabitants.

And it has become clear to me that the fossil fuel industry is anathema to democracy. We cannot continue to depend on fossil fuels and live in a country that calls itself a democracy. Here's an example: A poll has shown that two-thirds of Americans want our government to pass a binding climate treaty. But in 2013, the fossil fuel industry spent $326 million to persuade congressional Republicans and some Democrats to kill just such a treaty. At the local, state,

---

*Benjamin Franta, "Shell and Exxon's Secret 1980s Climate Change Warnings," **Guardian**, September 19, 2018.

national, and international levels, again and again, fossil fuel money pours in to stop any measure that seeks to limit their ability to maximize extraction and profit over the health and desires of the public. This shows how fossil fuels are undermining our democracy, subverting Americans' wishes.

Fossil fuels are a glaring example of corporate socialism in our country. Our government subsidizes them every year with $20 billion in subsidies and tax breaks. Forty-five percent of existing drilling and fracking would not be profitable without government support. That is what corporate socialism looks like, so let's not allow them to attack the Green New Deal as "socialist."

And on top of all these past crimes, they are destroying our future. All this so that a small number of fossil fuel industry executives and arms manufacturing CEOs, and the politicians they bribe and the banks and investment firms like JPMorgan Chase, Wells Fargo, Union, and BlackRock that underwrite them, will grow richer and richer.

They were certainly growing richer that week. When Trump assassinated the Iranian Quds Force commander, Qasem Soleimani, on that very Friday and the United States and the Middle East seemed braced for possible war, the price of oil soared, and as a result all the pending pipeline projects seemed doomed to succeed. It was a stomach-churning time on many levels.

The Teach-In

For our teach-in on how to hold the culprits accountable, we had a truly fierce foursome of females. Ellen Dorsey is a longtime human rights advocate who currently works as the executive director of the Wallace Global Fund, a private foundation focused on progressive social change in the fields of environment, democracy, human rights, and corporate accountability. Ellen has worked at the intersection of human rights and environment for years; she started Amnesty International's groundbreaking program that combined the two and, inspired by her previous work against apartheid in South Africa, was an early leader of the fossil fuel divestment movement in the United States.

Next, Tamara Toles O'Laughlin is the North American director at the global organization 350.org, where she is building a multiracial, multigenerational climate movement capable of holding our leaders accountable to science and to justice.

We also had Katie Redford, a human rights lawyer and activist who holds corporations accountable for human rights and environmental abuses around the world. Katie is the founding director of EarthRights International, which has represented local and indigenous communities fighting against corporate wrongdoing in places like Burma, Nigeria, Standing Rock, and Peru. Currently EarthRights is suing Exxon and

Suncor for their role in the climate crisis. Katie now leads the Equation Campaign, dedicated to stopping the expansion of fossil fuels, supporting movements in frontline communities, ensuring that climate activists have great lawyers, and keeping fossil fuels in the ground.

And speaking of frontline communities, our final guest was Janene Yazzie, a Diné (Navajo) woman who is the co-founder and CEO of Sixth World Solutions, which works with Diné communities to develop projects and policies that promote sustainability, environmental justice, and self-governance. She also serves as sustainable development program coordinator for the International Indian Treaty Council and co-convener of the Indigenous Peoples Major Group for Sustainable Development, engaged with the UN Sustainable Development Goals.

Ellen Dorsey started us off, saying, "It's clear that we live in a moment where the power of corporations and economic institutions outstrips the power of governments. That there is a global phenomenon of corporate capture of democracy, and what that means is that governments work on behalf of their donors. They work on behalf of the most powerful economic actors in the world and not in service of the public good. And Jane was right. The way this played out in the climate crisis is that environmental advocacy organizations were spending their time pushing for policy and policy change. And my sector, philanthropy, which funds groups, was putting money into that policy change, tens of millions of dollars, busing students down to D.C. to lobby for the public policy du jour. But we saw absolute failure to get any meaningful policy passed. And efforts to get a global deal at the Copenhagen UN Conference failed miserably. It was clear that we weren't going to get policy change.

Ellen Dorsey speaks.

"We weren't going to get a global deal, because the industry was spending tens of millions more than my sector in denying the science, promoting misinformation, lobbying for government inaction, lobbying for governments to continue subsidies to their industry as opposed to the alternatives. And so a scrappy band of students began to turn the target on the fossil fuel industry, and they took a play from the antiapartheid movement and said, 'We're going to call on our universities to divest from fossil fuels.' And they started calling on their universities to divest. And that spread very quickly. It spread to forty campuses overnight. It spread to the faith groups, to philanthropy, my sector, pushing us to divest our assets from fossil fuels because of the contradiction. How can you fund climate activism while your endowments are invested in fossil fuels? So what began as a scrappy little movement in 2011 has now exploded to a movement that is a full-blown global social movement.

"In 2014, the movement had hit the level of moving $52 billion in assets under management. Today, it's $12 trillion in assets under management. That's people power, and that's people power demanding that corporations are accountable to us and to the common good. And it's something that any of us can engage in."

Tamara O'Laughlin spoke next. "Ten years ago, we would have been laughed out of a room full of friends saying that we need to end fossil fuels, even though they've been the same enemy since day one. It's not just fossil fuels; it's their deep-pocket enablers, who live in legislatures and sometimes pretend to listen to us. It's talking about how we must focus on the insurance agencies that tell you 'no' to fire, flood, and crop insurance but 'yes' to insuring oil pipelines, thereby making sure that the climate crisis continues. Compressor stations, well pads, all of that oil infrastructure stuff that shows up in your community cannot pass 'go' unless it's sponsored by a bank and an insurer.

"But when we think about what we have to say 'no' to over the next ten years, we also have to talk about what are we building in its place? What are the kinds of things that will have to be built in the future so that we will feel very different than we do now, and who's the 'we' that we're talking about? Just to be clear, black, indigenous, and people of color have not been included in the 'we' until two minutes ago. And they've been doing the work for forty, fifty years, since before organizations came together to do this work."

"I want to kind of backtrack a little bit," Janene Yazzie said, "to go back to this question about who's to blame and how did we really get ourselves here? Because I always kind of cringe inside when we talk about going back to when our democracy worked. And I'm like, 'Well, what era was that? And what was happening to my people at that time?' But I think it's really important to point out this history, and this is what indigenous peoples have been trying to do for generations. Before Exxon even knew what they were doing, we've known and witnessed these changes they were bringing, because our territories, our lands, our resources, were being used through colonization and through just outright theft to feed the development that we are seeing that has caused so much damage and injustice and systemic oppression in our communities. This influence of money in politics is not a new issue in Washington. This government was structured off of white landholding men having all of the power and all of the authority to make decisions that affected all of us. I suppose they did try to create a democratic government that Benjamin Franklin structured based upon the Iroquois Confederacy. The confederacy did have a true egalitarian society, a true egalitarian governance structure that was rooted in and respectful of our relationship to our Mother Earth, to the natural world, to our ecosystems, what people conveniently call nature-based solutions now. But the white colonists did it wrong. They did it wrong because embedded in the development of their institutions of power, their institutions of governance, and economies was the exploitation of things that weren't considered valuable: people, women, plants that weren't valuable to the wealth, and to the power, and to the vision that those original white male landholders saw for themselves. But if I, as an indigenous woman, were to fight for our lands, I'd have a target put on my back because we're living in a society, in a First World country, that still targets and criminalizes indigenous and land rights defenders; that still covers up the epidemic of missing and murdered indigenous women in our territories.

"I heard this new term, 'climate grief.' People are feeling overwhelmed by all of these things that we're dealing with and all of the pressures to change so much in so little time. And I'm like, 'Join the club.' We've been dealing with this grief, with confronting our apocalypse, our genocide, for generations. And we're still here, still trying to give solutions that will help everyone, that will help all life-forms, because at the end of the day, I feel like we're often treated as we're just the pain in people's side because we're always harping about our rights. But our rights are what's going to actually help amplify and magnify the solutions that we bring to the table and their successes in preserving ecological biodiversity, but also in informing what true, just, egalitarian government structures look like.

So yes, we're going to harp about our rights, but we're harping on it because our rights are your rights. I am Diné. It means 'the human beings,' it means 'the people,' it means recognizing when we greet each other that you are the center of your universe, I am the center of mine, and together we live in this world as one. And that we live in this world, not to see what we can do with it to benefit ourselves, but how do we become community and contribute to community?"

The themes expressed by Janene are ones threaded throughout many of the indigenous speakers we've had at Fire Drill Fridays.

Because of the time I have spent in Canada at climate rallies, I am well aware of the central roles First Nations peoples have always played in the fight there for climate justice. Indigenous leaders are always present at the front of all rallies, the first to speak, their demands the first to be lifted by the entire Canadian climate movement, and they have a long history of putting their bodies in front of pipelines and stopping them. I have been at Standing Rock, in North Dakota, during the water protectors' siege against the Dakota Access Pipeline. But only after four months in D.C. did I truly understand how central indigenous peoples in all parts of the Americas have been to the fight against oil and gas pipelines, toxic dumps, and coal and uranium mining. And, like their Canadian counterparts, they have frequently won. The toxic perpetrators have been doing their digging, drilling, and dumping on indigenous lands before the rest of us knew there was a climate crisis looming. But in the United States, indigenous peoples' essential contribution to the climate struggles is far less known and much less appreciated. I hope that this book and Fire Drill Fridays will help change that.

When the time came for audience questions, the first questioner wanted to know why the insurance companies continue to invest in fossil fuel when it is clearly a failing sector with no long-term viability.

Ellen answered by saying, "Why the insurance companies are actually operating against their own interests by insuring infrastructure, even the health insurers, is a good question. If you look at the impacts of climate change, it is going to cost the insurers more money to pay out for damages. So we need to pressure them to understand that they need to take on the fossil fuel industry, because the fossil fuel industry is going to cost them money. We have to target them and campaign against the insurers. Tell our insurers, 'Do not underwrite new fossil fuel.'"

Greg from New York asked a question that roused everyone: "If fossil fuel companies actually paid for the damages they caused, what would be the best use of redistributing that money?"

"Funding the Green New Deal!" Tamara jumped in to say. "Over and over and over again. Funding the Green New Deal is not just a great road map; it is a thing that can be funded at the local, state, and federal level. It's a thing that can be used to make sure that what we're investing in is not what we're currently

CLIMATE JUSTICE

Katie Redford

"I'm the founding director of EarthRights, and our mission is to combine the power of law with the power of people to defend human rights in the environment. The only way to make these companies change is to make them pay for their dirty practices, because hitting them where it hurts—their bottom line—makes them actually listen. One way we do this at EarthRights is by litigating against fossil fuel corporations with the goal of holding them accountable and making them pay their fair share of the costs associated with climate change. EarthRights has a twenty-five-year history of suing oil companies. Not exclusively, we go after many kinds of corporations that are complicit in human rights and environmental abuses. Over the years, we have sued companies like Unocal, Shell, Chevron, and Chiquita, and not only have we sued them, but often we've won! We've won cases on behalf of some of the most vulnerable communities in the world, from Burma, to Nigeria and Peru, to

Standing Rock. And these are just a few of the cases that we've successfully litigated. So when people say that it's impossible to take on this industry, I say no it's not. It's possible, and we're doing it right now.

"The entire system needs to change so that the corporations who have benefited from being above the law for so long are now held to account and have to follow the same rules that we all do. We need to hold those wrong-doers accountable. They cannot get away

Katie Redford risks arrest after the rally.
Behind her is Sam Waterston.

with this anymore! So that's exactly what, at the end of the day, litigation really needs to be about. It's about punishing the perpetrators to make sure they don't keep doing it. It's about making sure that they internalize the costs of their operations that they have put on us for so long. At the end of the day, we must deter future abuse and make these destructive activities by these abusive industries unprofitable.

"And so what climate accountability litigation does is to say that your time has run out. It's time for you, the fossil fuel giants who caused the problem, knew about it, concealed it, lied about it, and then made billions off it—it's time for you to pay your fair share. That would be a small step toward justice, and that's what we're seeking. And that's what I call accountability."

invested in. And all of that stuff can happen today. There's a road map for it. And I'm not just saying that as someone who supports it wholeheartedly. I'm saying it as someone who recognizes that the thing that we're looking for is already built. We just need to invest in it. And that's beyond getting regulators to stop recklessly spending our money on stuff. We want governors to stop ignoring us, stop making backroom deals that don't care about our interests, and start funding the things that will pull jobs, human health, and infrastructure out of the gutter."

"The money we're spending on subsidies for the fossil fuel industry, it is absolutely criminal," added Ellen. "The International Monetary Fund put out a global report six months ago. Five point two trillion dollars in 2017 went from world governments', that means public, funds to the fossil fuel industry. Ten times what was spent on public education. That's criminal. How can we be using our tax dollars to subsidize this industry at this moment in time that is literally destroying the public good?

"There's one thing I think we have to be really careful about," cautioned Ellen. "Any policies being pushed by a government who, by and large, is still beholden to its donors and the industry is going to indemnify them. And so, for any legislation that's being proposed on climate, we have to stand vigilant and ensure they cannot be indemnified in that legislation."

A questioner from Florida asked, what about the

politicians who have fossil fuel companies in their pockets?

"Vote them out," said Janene. "We have so many people who should be in politics that don't run just because of the amount of corporate money and dark money behind our political systems. But run anyway. Build your base. Build your platforms and get things moving. Know the names of people who are in the pockets of these corporations. Hold them accountable to what they need to do, what they were voted in office to do, and then watchdog them and get them out of office."

A questioner from Brazil wanted to know how the fossil fuel industry is fighting back against climate activism.

Janene answered, "Laws that criminalize protests are being put forward in every state in the United States. They're called critical infrastructure laws. And this is a pattern we've seen all over the world. This isn't unique to the United States."

"The way they do it," said Tamara, "is by focusing on private property and by persuading state legislatures—and they even get labor behind it—that fossil fuel infrastructure is so critical that to even plan to stage a protest near it is to endanger national security and is considered a crime. They're making the cases tighter and tighter. And that's why we have to build an army between now and November that can register people to vote and make them understand whatever your issue is, climate has to be number one. You have to be sure that the politician you're voting for, whether it's a state legislator, a member of the board of supervisors, a sheriff, or a senator, or a congressman, or president, that they do what Ellen said. No corporate money. No fossil fuel money. That they divorce fossil fuels."

"We have to also remember that this is an industry that has not been toxified yet," added Katie. "I mean, it is critical that fossil fuel companies go the same way as big tobacco or the opioids. They absolutely are the same; it's the exact same playbook. And so we can use that playbook too because right now a politician would not take money from the opioid companies. That would be ridiculous. They would be shamed. And it needs to be the same way for fossil fuel corporations."

Louis, from Puerto Rico, asked, "What can we invest in to support fighting the climate emergency?"

"Localized solutions," answered Janene. "Especially in Puerto Rico, there are several examples of different communities building self-sufficiency and self-sustainability, localizing their economy. And indigenous peoples in Puerto Rico have been leading these solutions themselves in their own way."

"Yes, and change the models of ownership," added Ellen. "Our foundation is working with the Standing Rock Sioux tribe that we all know was the center of a fight against an oil pipeline. They're building a grid-scale wind farm that will bring in revenue. Majority owned by the tribe that private investors can support and bring in capital that will generate substantial revenue for the community. We as investors who must divest from fossil fuels can invest in this.

"There's a billion-plus people in the world today that don't have access to electricity," Ellen said. "They couldn't be reached with heavy, fossil fuel grid infrastructures. Now, with technology, those communities can be reached by leapfrogging fossil fuels with distributed, small, locally owned energy that allows children to study at night, to store vaccines twenty-four hours, to have women walk safely in their communities. The access to energy needs to be recognized as an international human right. We have the ability to reach those billion-plus people, and that's a justice issue." This spoke to what all my guests had underscored: the need for decentralized, locally controlled energy sources.

Ted texted in from Montana asking if the media is complicit in getting in the way of people knowing the full truth about the climate crisis.

Tamara replied, "They fail to tell the stories. Media has not jumped into the forefront of this conversation around teasing out what liability means, helping people to understand how we move from talking about the toxic problem you see where you live, to the bank that's sponsoring it, to the insurance company that makes it happen, that's megaphone-strategy-type work where you really have to tell the story in a way that helps people see themselves in it. And unfortunately, the climate crisis is delivering us a lot of stories. So if the media's going to show up, it's going to have to start telling stories of people losing their land, losing their lives, losing their relationships, and their communities being destroyed if we're going to move the needle in the time that we have."

"But shout out to the great journalists at Inside-Climate News and others that exposed the Exxon story," Ellen added. "We need to support the independent journalism and media that do still bring us those stories."

"Thanks for watching," I concluded. "And I hope you agree with me that we really drilled down into some critical information tonight. See you next time." It was my own little sort-of-joke that I always closed the teach-ins the same way I used to close my **Workout** videos. "See you next time."

The Rally

Sam Waterston had returned for a second Fire Drill rally, bringing his sister Ellen and our mutual friend Judith Bruce, all of whom were planning to engage in civil disobedience. He introduced the first speaker, Janene Yazzie.

"The lands that I come from know too well who's to blame for the travesty and the calamity that we're in," Janene told the crowd. "I grew up in the territories

that have been im-
pacted by uranium
mining and its legacy
since those activities
took place on our
lands. The legacy that
is found in the bodies
of our unborn chil-
dren who enter this
world with uranium
in them at the level of
adult miners. I come
from the lands that

Janene Yazzie speaks.

are ravaged by coal mining and natural gas exploitation and by all of the residual impacts of systemic
oppression and injustice that comes from being a resource colony that has been exploited and used for the
benefit of people who don't know the true cost of what
they've become dependent upon. One of those true
costs lies in the shortening of the lives of our relatives.

"Even today, a dear friend and mentor, Robert Toki,
who dedicated his entire life to social, racial, economic, and environmental justice, is being put back
into our Mother Earth's arms in Flagstaff, Arizona,
because he lost his battle on Monday to stomach cancer. He is one of a long list of loved ones who have
gone unknown but who, like Greta, have been working since they were young to shed light on this issue
that we are facing. In his last interview, his last words

that he left with us, Robert said, 'If we can all remember that we are interconnected, we will be okay.'

"So when indigenous peoples are coming to you, trying to make space in these areas, and in these conversations, we do so because our relatives are dying in real time; because every aspect of exploitation on our land is connected to the subjugation of our children and our women. There's always been a connection between the institutions that have supported and built their power off fossil fuel extraction and what is going on with the prison pipeline, the separation of children and families at the borders, the disappearance of our women, and the lack of access to clean drinking water. Here in a First World country, residents in Flint are still feeling the same effects that we feel on our reservations with not getting enough resources to ensure that their children can have this basic necessity for life. And this is being multiplied all around our country. It is unacceptable. And it has been the lack of awareness over generations, and the way that these institutions were founded upon white heteropatriarchy, that has left us blind and unable to build the solutions that we need to. But we say, 'No more!'"

The next speaker, Donna Chavis, a member of the Lumbee tribe of North Carolina, is the senior fossil fuels campaigner with Friends of the Earth and a recognized leader in social and environmental justice and was a member of the planning committee of the first National People of Color Leadership Summit in

1991, which developed the Principles of Environmental Justice.

"I bring you some good news first. I received notice yesterday that at the end of the year Duke Energy reached a settlement with the communities and the environmental groups in North Carolina for the disposal of eighty million tons of coal ash. We have spent most of the last decade fighting Duke Energy over coal ash, because when their coal ash spill into the Dan River happened in 2014, the Band-Aid was torn off all the coal ash dumps that were all over the country but especially in North Carolina. All Duke wanted to do was cap them and leave them in place, which meant the water for those communities was going to continue to be ruined. Families are dying. There was a death just last month, right before Christmas, of one of the long-term organizers against the coal ash. And so I ask you with great joy and celebration to raise up a cry for all the organizers, all of the climate activists, who fought against Duke Energy, the largest energy company in the United States, and won. And won!

"Now, as has been said already, we are all in this together, but so are they. Duke Energy has joined with Dominion Energy, and at the same time as the Dan River spill happened, these two largest energy companies in the United States were planning the six-hundred-mile Atlantic Coast natural gas pipeline running from West Virginia supposedly to North Carolina.

"I'm here today to give you some of the information about this pipeline, because it is a huge deal. In fact, on February 24 the opening arguments before the Supreme Court will be heard in the case that challenges the Fourth Circuit's decision that the U.S. Forest Service did not have the right to permit this pipeline to cross the Appalachian Trail. Imagine, this would be the first time that a pipeline has crossed a national park. It is a climate, human rights, and environmental disaster. If it is completed, it will carry enough gas to generate over sixty million metric tons of climate pollution per year.

"Along the route of the pipeline, we have at least nineteen tribal nations that are being impacted. In North Carolina alone, we have over thirty thousand indigenous peoples that are impacted. And yet you know what they said? 'There's no disproportional impact.'

"We want to protect our communities and protect our health and our history. Energy Justice North Carolina is a collaboration of environmental groups and citizens that are fighting the infrastructure of the energy companies, not just through demonstrations, not just through regulations, but through getting banks and legislators to do what is right. In fact, Bank of America dropped Duke's rating because they said that they have oversold how easy it would be to get that law passed. So that was one time when the bank stepped up."

Next, Tamara took to the stage. "My name's Tamara O'Laughlin, and I'm here to make it real plain. The climate crisis is here now, but it's a crisis of courage. Do you have the courage to tell the truth this afternoon?"

And Tamara then told the crowd about "10 Steps for the Next Administration's First 10 Days," which you can learn more about at ClimatePresident.org. "We don't know who the president's going to be. Okay? We're going to do our best. But we don't know. But these are ten things that no matter who it is, even if it's him, that we have to make happen. Because we won't be able to make it in the next decade if we don't make these things happen. So let's keep these in our mind. I mean, that's a lot of work we got to do."

Tamara Toles O'Laughlin speaks.

COAL

Veronica Coptis

"I am the director of the Center for Coalfield Justice and, just a four-hour short drive from here in D.C., we work with communities whose land, water, air, and people are being exploited for private profits by major fossil fuel companies. But the identity I'm most proud of is being a mother of a spitfire three-year-old and a precious two-month-old, who are back home with my husband right now. But also, being a mother on the front lines is full of terror, worry, and anxiety. I personally grew up in the shadow of the largest underground coal-mining facility in North America. To put it in some perspective, it's larger than the island of Manhattan, and, by the way, it is still producing twenty to thirty million tons of coal a year. So we may think coal is dead, but it is nowhere near dead and still destroying our communities. I currently live surrounded by fourteen gas well pads within a mile of my home, and I'm not able to trust my public drinking water to make formula for my baby. I'm fortunate enough to be able to buy an alternative water supply, but how many mothers in our community cannot? This is just my story. And hundreds of mothers are dealing with real health impacts to their children, of nosebleeds, rashes, rare cancers, and even having to bury their own children right now. And there's been lots of hopeless speakers and I'm very angry. Think back when your mother was angry at you. There's nothing more dangerous than an army of angry mothers demanding action for our children right now. Frequently, when I travel to big metro places and more progressive-leaning areas, I get asked the question, 'If all of these impacts are happening in your community, why aren't more people resisting and fighting back and stopping this?' Well, people are resisting. They have been resisting for centuries in our communities, but the coal and gas companies have been monopolizing our economies and corrupting our political systems at the local, state, and federal level and oppressing any ounce of hope that we have the power to win. But that is changing. And our organization at the Center for Coalfield Justice is shifting the mentality, and culture, and hearts, and minds of the places that many of you might consider Trump country, but it is far from it. And we're doing this by investing in the residents who live there and leading by examples. Our team is primarily composed of young folks and community members who have lived in our region for decades and are also directly impacted. And I can tell you, the easiest way to move a community is to pay the people who are from there to organize their own neighbors. But that comes at a risk. By just speaking out, by me just coming to this rally today, puts a target on our faces, and billion-dollar energy companies will wage character-assassination campaigns against individuals. Threats of violence on you and even for some folks the risk of their economic stability. But we are doing it despite all those threats. At the Center for Coalfield Justice, we have hundreds of brave members who are banding together and starting to show that we actually can create change in our own communities, and we are doing this by demanding that the governor of Pennsylvania finally put money into researching the fracking and its serious health impacts. We are doing this by suing one of the oldest and largest coal companies in the country, and we stopped them from undermining streams in a state park, costing them roughly $10 million in production.

"And we're educating and registering voters to get engaged so that we can now have a stronger voice in our political system. But beyond these Fire Drill Fridays, I really hope you take the privilege that it is to get arrested here in D.C. and know you're likely going to walk away, and work with community-based organizations that are taking these direct actions on the front lines as well."

10 STEPS FOR THE
NEXT ADMINISTRATION'S FIRST 10 DAYS

In early December, a consortium of policy experts released a plan, supported by five hundred organizations, that urges the next president to take ten executive actions, starting on day one, to confront the climate emergency without waiting for Congress.

By the time of the inauguration, we will have less than ten years left to reduce fossil fuel emissions by half. We are saying to the next president, "We have ten years; you have ten days." Ten executive actions to do in ten days to make history and change course away from the climate cliff.

1. Declare a national climate emergency under the National Emergencies Act.

2. Keep fossil fuels in the ground.

3. Stop fossil fuel exports and infrastructure approvals.

4. Shift financial flows from fossil fuels to climate solutions.

5. Use the Clean Air Act to set a science-based national pollution cap for greenhouse pollutants. Then use all Clean Air Act programs to drive emissions toward zero economy-wide.

6. Power the electricity sector with 100 percent clean and renewable energy by 2030 and promote energy democracy.

7. Launch a just transition to protect our communities, workers, and economy.

8. Advance climate justice: direct federal agencies to assess and mitigate environmental harms to disproportionately impacted indigenous peoples, people and communities of color, and low-wealth communities.

9. Make polluters pay: investigate and prosecute fossil fuel polluters for the damages they have caused. Commit to veto all legislation that grants legal immunity for polluters, undermines existing environmental laws, or advances false solutions.

10. Rejoin the Paris Agreement and lead with science-based commitments that ensure that the United States, as the world's largest cumulative historical emitter, contributes its fair share and advances climate justice.

The civil disobedience blocking the intersection between the Capitol and the Supreme Court was full of energy, diversity, and notable leaders. Ka Hsaw Wa was

Sam Waterston during arrests.

among them. An environmental and human rights activist from Myanmar/Burma who has done dangerous, truly life-risking investigations there about oil company abuses, he was also risking arrest, which is impressive, because Ka Hsaw Wa had been jailed and tortured for his student activism in Burma in the late 1980s and doesn't take arrests lightly. In the early 1990s, he discovered that killings, rape, torture, forced labor, and relocation of his people's villages were connected to the exploitation of natural resources in the name of development and that fossil fuel industries, including the U.S. oil giant Unocal, were complicit in the abuses. Ka Hsaw Wa, incidentally, is married to Katie Redford, one of the day's speakers, and together they have been bringing lawsuits against corporate human rights abusers for twenty-five years, including against oil giants like Exxon and Suncor for their role in the climate crisis. The couple were both arrested that day, alongside their daughter, Alexis. After the arrest, Ka Hsaw Wa explained that he "joined the arrest to show people back home [in Burma] that

GAS/FRACKING

Josh Fox

Josh Fox is a documentary filmmaker.

"I spent the last ten years of my life seeing the effects of the fossil fuel industry on the ground, on people, on the frontline communities. And I can tell you from up-close experience that this industry kills people. And it kills children. And it does so with impunity, and then it lies about it all across this country and all across this world. The fossil fuel industry goes into people's backyards and destroys those places. It pollutes the water, it contaminates the air, it causes a health crisis. And it destroys communities at the very root, and then lies about it, and lies about it, and lies about it, and lies about it.

"What does it do to people, the fracking industry? Well, I can tell you about sitting at people's kitchen table and hearing them say, 'I have cancer. My children have cancer. You see that dark spot on my MRI. That's brain damage.' Like Amy Ellsworth in Fort

Lupton, Colorado. Her water was so flammable because it was contaminated by the fracking industry that she could light it on fire and it was causing little mini explosions in her bathroom sink. She was showering in the dark, because she was afraid a spark from her lightbulb in her bathroom would blow her across the street in the middle of a shower. Or Terry Greenwood in southwestern Pennsylvania, who died of glioblastoma brain cancer. And I remember him wrapping a calf in plastic and stuffing it down into a meat freezer. It was stillborn, born blind with pure white eyes. He said, 'I'm going to find out what killed my calves,' because they were drinking out of the frack pond up the hill. We buried Terry before he got a chance to autopsy those calves. Or Norma Fiorentino in Dimock, Pennsylvania. Her water well exploded on New Year's Day. And everyone came home from church and the concrete casing was headed out all throughout the yard and people started to compare stories. They said, 'My laundry water turned black. My children are getting sick. My animals are dying.' This is what the fossil fuel industry does in America. You can go to state parks in West Virginia, camp out, listen to a babbling brook; wake up the next morning, you see it's running bright orange from coal mine runoff.

You can go to the Amazon where Shell and Chevron are polluting the most biodiverse atmosphere for fish in the world. And there's oil spills. And villages of people in the Amazon are having to buy canned tuna because they can't eat the fish, because it's so toxic and it's so polluted. And you can see the ravages of climate change in Paradise, California, where people are watching their neighbors exploding, burning alive in their cars running to parking lots where they took shelter because the only thing that didn't burn in that part of suburban America were the parking lots. You can go to the Virgin Islands and see people whose houses were demolished, and they're burying their own pets, again, because the floodwaters unearthed their pets' graves. This is what the fossil fuel industry does. And you can go to Standing Rock as I did. And you can see those people in the water, the water protectors, led by that movement for indigenous sovereignty and joined by people from all colors all around this nation. People from Ferguson who are getting teargassed, Black Lives Matter people standing there in the water, the freezing November waters in North Dakota, saying to the police, 'We love you. What are you going to say when your child wakes up the next morning and says, "Daddy, can I have a drink of water?" Are you going to say you were standing here with us? Are you going to tell your child that you were holding a gun to my head for protecting the source of water for eighteen million people and then the whole line was pepper sprayed, maced right in front of you?'

"So, I ask you, how many of you have actually been to a fracking site? You're wrong. You're at one right now. This is the biggest fracking site in the United States of America. And they're not injecting toxic water and pollutants as they do in the frack zone. They're injecting toxic dollars by the billions. And they're telling your politicians to ignore your concerns. They're telling the politicians to ignore the people who are dying all across this nation. They're telling the politicians, 'Don't listen to those stories. Don't listen to those people. Just keep taking the money.' Well, what do we have to do? We're in 2020. Rex Tillerson, that doughy, dimpled denier, CEO of ExxonMobil, becomes secretary of state, only to be replaced by Mike Pompeo, another fossil fuel executive. ExxonMobil is running the State Department. We have to vote out Donald Trump.

"That's the number one thing. We have to get real on this administration. We have to put them in jail. If someone comes into your house and burns down your house, they go

to jail. If some-
one goes into your
house and steals
all your most pre-
cious things, they
go to jail. The fos-
sil fuel executives
must be going to
jail. And there are
Democratic can-
didates right now

Josh Fox speaks.

who are saying, 'We are going to jail the fossil fuel executives.' Are you with me? We have
to stop fracking. And we have to stop fracking the capital. We can take back our government. We have to mobilize in this election.
Thank you so much."

RESIST

Janet Redman

Janet Redman is the climate director at Greenpeace USA.

"This fall, Greenpeace took the fight right to the heart of the fossil fuel industry. We blocked the Houston

Janet Redman speaks.

Ship Channel for a day, stopping oil tankers from entering, and, yeah, this single act didn't bring the fossil fuel industry to its knees, but what I was excited about was it really showed the CEOs of oil companies that we're willing to take them on even in the heart of their power. And for a day the channel was quiet. It was transformed into a vibrant waterway full of life again. And I swear to God there was a rainbow that came out in the sky. I spent last week celebrating the holidays with my nephews in Florida. And after they'd gone to bed, I had some time to sit in the living room with the Christmas tree lights lighting up the room and think about the new year is coming and what kind of resolutions I'm going to make. What am I going to recommit to inspired by this abundance of incredible activism that happens every day? I'd like to publicly resolve here that Greenpeace will take on the fossil fuel industry this year with renewed vigor. We'll work our butts off to topple the political-tech industry, and financial pillars that continue to prop it up. We recommit to aligning ourselves with frontline fighters, building bridges across the movements to grow our collective power, and we'll remember to keep hope and joy at the center of our struggle to shift the collective imagination about what's politically possible in this country.

"I believe that if climate criminals can't quench our thirst for justice, if they can't poison our love for one another, then they can't break our resistance, nor can they extinguish our vision for a future freed from fossil fuels. I'm anxious to start joining forces with you all for the fight this year and far, far beyond."

The group looks on as activists risk arrest. Jane had already been arrested too many times to be able to risk arrest again without considerable jail time. She stands with Jerome Foster.

climate is not just an American issue but that it is an important issue for the world."

What Can I Do?

A good place to start holding fossil fuel corporations accountable is ensuring that our own money, as well as that of institutions we have influence on, does not support them and instead supports real climate solutions. By moving money out of fossil fuels, we align our investments with our values while also weakening the industry that has refused to act on climate change, lied to the public, and blocked remedies.

First, you can "divest-invest," which means moving your money out of supporting fossil fuels and into companies supporting climate solutions. Many individuals' retirement and savings accounts are invested in index funds that are considered passive funds, meaning that they are expected to produce returns consistent with broader market indexes. These passive funds generally do not evaluate corporate behavior and certainly don't screen for fossil fuel investments. Check out Fossil Free Funds (fossilfreefunds.org) to get a list of high-performing mutual funds to invest in that exclude companies that violate fundamental environmental justice, human rights, and fair governance standards.

Second, many of the organizations and institutions that you are affiliated with have investments. Determine whether they are already a target of divestment campaigns and join in. If not, you and other like-minded members of those communities can call for them to divest from fossil fuels and invest at least 5 percent of their portfolios in climate solutions. These institutions respond to sustained pressure from student groups, academics, alumni, current and retired workers, clergy, parishioners, and others. You also can campaign to force your union pension, local government, city, or state to divest-invest. To find out more about how to plug into a divest-invest institutional campaign, visit DivestInvest (divestinvest.org) and GreenFaith.

Third, take fossil fuel money out of your political system! Make politicians in your area take a #NoFossilFuelMoney pledge. Taking the pledge means that a candidate will pledge not to accept any contributions over $200 from the PACs, lobbyists, or SEC-named executives of fossil fuel companies. You can find out about how to get involved at NoFossilFuelMoney.org. If you live outside the United States and your politicians are serving the interests of energy companies over your community, start your own drive to demand that politicians represent you, not fossil fuel companies. You can also help stop subsidies by adding your voice at Stop Funding Fossils (stopfundingfossils.org).

Finally, if you manage the investment portfolio at a company, organization, or institution, visit the DivestInvest.org guide to establish a six-step process for divesting your funds from fossil fuels and investing them in fossil-free companies. The Carbon Underground has a list of the world's top two hundred companies ranked by the carbon content of their proven fossil fuel reserves. Check Fossil Free Indexes (fossilfreeindexes.com) to see if any of your money is tied to these companies. The guide has clear criteria for investments as well as case studies that show how to divest from fossil fuels and invest in equitable, clean solutions. And don't forget insurance companies, a middle entity, that are now refusing to insure coastal properties, and other areas, at risk because of climate change. Demand that these companies also stop insuring fossil fuel projects that would not be economically viable without subsidies, insurance, and private financing.

Finance is an incredible lever for holding corporations accountable for their social and environmental impacts. It is no longer acceptable to own fossil fuels, finance fossil fuels, or insure fossil fuels. Make it known that if you own fossil fuels, you own climate change. Demand that they divest them now and instead invest in our future!

Jane hugs Naomi as she is released.

CHAPTER SIXTEEN

Stop the Money Pipeline

I was a parish priest for ten years, and I found that if you really want to get at the heart of what people's values are, you don't stand up and say, "Let's have a conversation about values." You talk about money, and you get at the values very quickly. And if people are invested in the fossil fuel sector, they own the climate crisis.

—THE REVEREND FLETCHER HARPER,
Episcopalian priest and executive director of
GreenFaith, a global multi-faith climate organization

I started packing that week of January 10. It made me sad. The almost four months in D.C. were the first time I had ever removed myself almost entirely from the trajectory of my life to take up an existential challenge with people I didn't know. I didn't ask anybody's permission and focused exclusively on something that had originated in my own heart. I knew that because of the activists who had helped it happen, it was making a difference in the broader scheme of things. Despite the frightening subjects I was learning about, despite being surrounded by and having to interact with strangers, which has always been hard for me, and despite the cold (which I hate), there have been few times in my long life when I have been happier. Now it was ending.

Tensions and talk of war were roiling the Middle East, the price of oil was ominously elevated, but on Wednesday of that week Annie Leonard had organized a two-and-a-half-hour Skype meeting with activists in California with whom I would be working over the next six months to conduct Fire Drill Fridays in my home state, while I filmed the final season of **Grace and Frankie**. After that, Annie and I would embark on a national Get Out the Climate Vote tour from August until post-inauguration. We'd also be helping build out the growing Fire Drill Friday movement. By this writing, seventeen thousand people have signed up on our website saying they wanted to hold rallies where they lived, and we wanted to increase that number. After November's election, we'd mobilize people around the "10 Steps for the Next Administration's First 10 Days." No matter who is elected, it will take the entire environmental movement to build the army that would make the plan a reality.

One by one the activists and experts introduced themselves, described the work they were doing around fossil fuels, and suggested various topics for our rallies and teach-ins. Annie invited Aileen Getty to join the meeting, and I was surprised to learn from her that California is the second-largest oil producer in the United States, after Texas. She explained how that made California an important battleground in the fight to stop fossil fuels.

After that meeting my sadness at leaving D.C. lifted. The first Fire Drill Friday in Los Angeles was less than a month away, and California was the place we needed to be. Our new, and apparently progressive, governor, Gavin Newsom, welcomed Fire Drill Fridays. I saw that we were entering another phase of building a movement, one where we grow bigger crowds, get people in the streets. Now that our base was California, I could get a lot more celebrities to come to the rallies and the demonstrations. With their help, plus the regularity of the actions, we were accelerating the movement toward change.

Meanwhile, in D.C., my wonderful lawyer, Mara Verheyden-Hilliard, had been negotiating with the D.C. prosecutor's office to avoid my having to return to the capital for my hearing in January, or to do community service in D.C., due to my four arrests. She succeeded in reducing all the potential penalties to simply my not being able to do civil disobedience or get arrested for three months . . . anywhere in the United States. I was once again reminded of the privilege afforded me because I am white and famous and can hire a fine lawyer.

The Teach-In

I was very excited that Thursday evening because one of my guests was Bill McKibben, the strong central pillar in the fight against fossil fuels and one of the first people I called for advice when I was starting Fire Drill Fridays. He's a deep strategic thinker and beautiful writer. One of those writers who, like Naomi Klein, puts things in a way that breaks through your protective armor, changing how you see things. It was he who launched the divestment movement and founded 350.org, a movement to bring renewable energy cheaply to everyone. Our teach-in and the rally the next day represented the "soft launch" of his new movement strategy: Stop the money pipeline. Protest against JPMorgan Chase as well as the insurance companies and the money management funds who make fossil fuels' heinous destruction possible.

My other guest was Eriel Tchekwie Deranger. I had met Eriel two years before when I'd visited the tar sands in Alberta, Canada. Eriel is a Dënesųłiné woman, a member of the Athabasca Chipewyan First

Nation in northern Alberta. She comes from a family of advocates who all fight for the recognition, sovereignty, and autonomy of their indigenous lands in what is now known as Treaty 8 Canada.

"Bill, let me start with you. Tell us about this movement that we are going to launch this month."

"So this is a crux moment, and at crux moments you've got to roll the dice and go big. One of the things that's going to happen in 2020 is a powerful focus on the financial backing of the fossil fuel industry. This is not new. People have been working on finances for a long time, and a lot of it began with indigenous communities taking on the people who are financing those crazy mining projects. And we've built this vast divestment movement over the last decade. Naomi Klein helped start it all. We're up to $12 trillion worth of endowments and portfolios that have decided to divest from coal, and oil, and gas. But now, now there's going to be a concerted movement-wide effort to go after the banks and the asset managers and the insurance companies that are the funders of this carbon bomb.

"We're launching tonight a new website, StopTheMoneyPipeline.com. And we need people to go there to sign up so that they'll be able to be activated in the course of the next few months. I know right now that all over the country and all over the world there are way more people who want to do something than feel like they have something to do, okay? We

Eriel Tchekwie Deranger, Bill McKibben, and Annie Leonard with Jane at the very last teach-in at the Greenpeace office.

finally reached the moment where people are, 'This is not the world I want for my kids. This is not the world I want for me.' And, 'But what do I do?'

"Well, maybe, and there's no guarantees here, but maybe we can figure out how to pressure the tiny handful of giant financial institutions that are the financial lifeline for fossil fuels. Take, for example, the single biggest lender to the fossil fuel industry, JPMorgan Chase. JPMorgan Chase, over the last three years, has lent $196 billion into this industry. If Exxon is a carbon major, so is JPMorgan Chase. They didn't need Donald Trump to pull us out of the Paris climate accords. They were sabotaging it from the moment that it was signed. And they're followed by Citibank, and Bank of America, and Wells Fargo. It's BlackRock in the asset managing world.

It's the insurance companies that instead of analyzing risk and helping us take care of the planet are pouring money into this system, further destabilizing it. Now, truly, it is daunting to take on the sort of central pillars of global capital, okay? These people are huge, and strong, and powerful. On the other hand, if we can get them to begin to bend, the effects will be rapid and global. Because all the money in the world is concentrated in a few places, New York, London, Shanghai, a few spots.

"And so, in the course of the spring, we're going to be asking people to be out in the lobbies of their local banks bringing this message. We're going to be asking people when the right date comes to be cutting up their credit cards at the same time. We're going to be asking people to pull their money. Because if you have your money in one of these banks, your money is not sitting there. It's being sent to Houston to hand it over to Chevron. It's being given to precisely the people who are doing precisely the kind of things that people like Eriel has been fighting forever, so I'm not telling you this is going to be an easy fight. It's not going to be. But it is the opening up of this work that people began, especially in indigenous communities, about a decade ago. And after the fight over the Dakota Access Pipeline, there was a big flowering of this work. Things like Mazaska Talks, 'money talks' in Lakota language, for example. And now belatedly but happily, across the environmental movement, everybody's coming together. If you go look at

StopTheMoneyPipeline.com, scroll down to the bottom and you'll see the logos of every major environmental group in the country now saying, 'We're going to take these guys on. Yeah, we know it's not easy. Yeah, we know they're powerful. Yeah—but we're going to try.' So we'll see. It's going to be an exciting year. StopTheMoneyPipeline.com.

"We have some experience thanks to people like Eriel in stopping oil pipelines. That's some of what we've actually kept from happening. Now we're going to stop this money. We're going to plug it up. We've worked hard, and we've done a lot of good against the oil companies. Shell, this year in their annual report, said that divestment had become a material risk to its business. That made me happy. Their business is a material risk to planet earth. Anything we could do to hobble them is a good thing. But Shell and Exxon and the others, they're going to fight to the last bridge because they know how to do one thing in the world, dig stuff up and burn it, okay? Chase, they'll lend to anybody. Whatever. Maybe they could even lend it to somebody useful like solar panels and wind turbines and things. Because we need money to bring those things quickly to the floor. So that's the thought. That's the plan. See you in jail."

"When you talk about Shell," Eriel said to Bill, "I think about some of the stuff I was doing ten years ago. I was working with the Rainforest Action Network, and we put together divestment tours in the U.K., where we heavily targeted Shell, Total, Norway's pension funds.

"The first year that we went, I spoke about the impacts in my community, a tar-sands-impacted community. We were fighting two projects, the expansion of the Jackpine Mine as well as the proposal for a brand-new mine called the Pierre River. And I came to bring evidence to the shareholder meeting, and we were literally laughed at and dismissed. I wasn't even addressed. The attitude was 'Don't worry about this person. They don't know what they're talking about.' The second year that I came back, they didn't laugh. They said, 'Now, thank you for your concerns, but you're wasting your time.' The third year that we came back, they asked us for a meeting.

"And now they're talking about how divestment has impacted their corporations over the years. And the reality is divestment works. When people stand up, and they continue to hold the line, it actually works. When you have people that are speaking real, lived experiences on the ground leading those campaigns, it works. But we have to also be talking about what are we asking them to reinvest in. It's not just talking about solar panels and wind turbines, but what are the regenerative economies that we're asking for the future? What are the economies that we're going to need in order to actually restore our relationships with the natural world, and uphold not just fundamental human rights and the rights of the environment but the fundamental rights of indigenous communities that are at the forefronts of extractivism and the forefronts of the impacts of climate change?"

Eriel then spoke passionately about the battle she and her family are fighting right now. They live in northern Alberta surrounded by caribou and bison in a critical habitat for endangered species. JPMorgan Chase is about to fund the largest tar sands mine ever proposed just seventeen kilometers from her community. Teck's Frontier project would be twenty-nine thousand square kilometers, twice the size of the city of Vancouver, and capable of creating six megatons of emissions annually. To make extraction easier, the mining company planned to remove what the industry calls overburden: the boreal forest, peat lands, wetlands, and soil, things that are essential to climate stabilization and to the livelihood of Eriel and her family.

(The very week I was writing this chapter, Teck withdrew its application for this Frontier megamine, marking a historic victory for the indigenous peoples and their supporters who had been blocking rail lines and bus terminals, basically shutting down transportation and occupying public spaces. Nonviolent civil disobedience, in other words.)

Bill said, "When we started the fight against the Keystone Pipeline, they were planning to quadruple production up there. We haven't shut them down, but that expansion is basically ground to a halt because there's no pipelines to get the stuff out and because people are pulling money out now instead of putting it in. It is amazing to think where we've come. I mean, when we came to Washington in 2011 to start the fight against the Keystone Pipeline, people said there was no chance. **National Journal**, a kind of trade paper here in D.C., polled its three hundred energy insiders, and 93 percent of them said Trans-Canada would have their permit by the end of 2011, okay? Then twelve hundred and some people, including a number in this room, went to jail, and it was the biggest civil disobedience action about anything in this country in a long time, and it started this ball rolling. It picked up on the work that people had been doing up in Canada, and with our colleagues out in the Midwest and all over the place, and then suddenly people are like, 'Yeah. We can take on big oil.' And so now every fracking well, every LNG [liquefied natural gas] terminal, everything gets fought."

Naomi Klein was in the audience at this teach-in, and she said, "I want to bring it back to something Jane said at the beginning about how the climate movement and the antiwar movement may need to be uniting once again. And, obviously, all of these issues are intensely interrelated because it's not a coincidence that so many of the conflicts are where the oil happens to be. So my question is for Eriel, because the first major tar sands boom that we saw was in 2003, after the invasion of Iraq. So can you talk about the interrelationship between war, oil price, and what's happening in your people's territory?"

DIVESTMENT WORKS

Bill McKibben

"When Naomi and I were first sort of thinking about doing this, we drew on the inspiration from the antiapartheid movement a generation ago. When Nelson Mandela got out of jail, one of his first trips was to the U.S. And he didn't go first to the White House. He went first to California, to Berkeley, to give a speech to students there, kind of representing students across America, to thank them for the work that they'd done to pressure divestment. South Africans liberated themselves. But as he said, 'We couldn't have done it on our own. We needed people bringing pressure from everywhere.' And divestment was a great way to do it. In the fossil fuel realm when we started, I think Naomi and I thought that the first task would be to sort of take away the social license of these companies. To get institutions to begin to say, 'We don't want to be partnered with you, because you're wrecking the planet and that's wrong.' And that worked. I mean, all the people who started it first were all the people you'd kind of expect. Small colleges and religious denominations and so on and so forth. And it was great. But as time went on, two things happened. One, people began to understand that these companies were also actually losing; they weren't good investments, because they have a great technological challenge coming from a better technology, sun and wind, and because we're going to hold them liable for the damage that they've done. So people who divested were making better money than people who didn't. And that began to add to the thing. And people began to see that it was effective.

"So when Peabody coal went bankrupt a few years ago, the biggest coal miner in the world, the government makes you list the reasons that you went bankrupt and the things that went wrong. And one of the reasons they listed for why they went under was that this divestment movement had made it almost impossible to raise capital. Last year, in March, at the big meeting of coal executives in Houston, **Politico** ran a story where they just one after another were saying, 'We can no longer raise money for expansion.' We've done a great job taking on coal. Coal is still an important fight to be had around the world. But now we have to bring just as much pressure on oil and gas and create the same kind of trouble. And it's happening. That's why I said it was such a moment this year when Shell said that this had become a material risk to its business. Look, four or five years ago, Shell was busily trying to go drill in the Arctic. Remember that big drilling rig they had in the harbor in Seattle? And thousands of people—we called them kayactivists—were blocking the harbor.

"Goldman Sachs, two weeks ago, rewrote their policies under a lot of pressure. They started talking about how they're not going to finance coal, and they're not going to finance drilling in the Arctic National Wildlife Refuge, which was a big victory for the Gwich'in people who have been fighting hard up there. I mean, let's be clear for a moment. It's an unbelievably hard fight to get one bank to say, 'We're not going to fund people to go drill in a wildlife refuge,' okay? I mean, if you even think about it for a minute. It's a wildlife refuge. Of course, you don't go drill for oil in it. But we're probably almost certainly going to, and almost certainly some of these banks, unless we really get in the way, are going to finance it because all they think about is, 'Can we make a lot of money in a little time right now?' They're not thinking like normal people would think."

Naomi Klein speaks.

"That's a very good question," Eriel said. "It's one that I don't talk about a lot, because it's really complex. We saw the price of oil rising in the early mid-'90s. But the Iraqi War really created this massive spike. And until that point, we did not see the expansion of Alberta's oil sands. It almost doubled overnight when we saw the price spike. And so there is absolutely a direct correlation between war, which causes this, particularly in the Middle East, where we have the richest oil deposits in the world. So when we destabilize those oil markets and cause the price to rise, because we can't access this sweet crude oil they have, it allows us to justify the expansion of bottom-of-the-barrel dirty oil, or heavy oil projects like the Bakken oil field, like Arctic drilling, like Alberta's tar sands, like the Permian oil basin. It allows us to justify these expansions of things that are in our own territory. It's more about this geopolitical struggle to maintain superpower status while destabilizing other places. Because America has re-created this sort of petro-state economy. And then the Middle East ends up landing on top of the richest oil deposit in the world. And of course, we have this struggle, and it gets really complex and dirty. But what are we seeing right now? The price of oil has been crashing, crashing, crashing consistently since 2011. And people are pulling out of things like the Bakken, and the Permian, and the Arctic, and our superpower status is becoming diminished in this place."

In adjourning that final teach-in, Annie started by thanking all of the "Fire Drillers" who had been tuning in every single week. "We're averaging over fifty thousand views of people watching this. We've now reached over seven million people if you count all of our social media with this. It is absolutely incredible what has happened. We've had over five hundred people join us to put their bodies on the line with civil disobedience and risk arrest. And I understand we're going to have over three hundred more tomorrow." Annie explained what was planned for California and for the Fire Drill Fridays rollout.

"We're going to be working with all of you who want to do Fire Drill Fridays in your community. We got so many requests. Just the other day, I got a request from somebody in Humboldt and somebody in

Philadelphia that both said, 'I want to do a Fire Drill in my community.' And I said, 'Great. We're going to develop tool kits and guides.' I asked, 'What do you need?' The one in Humboldt said, 'Can I just use your name? I'm ready. I'm going.' The one in Philadelphia said, 'I've never done something like this. How do I start?' We're here to help you, whichever side of that spectrum you're on, we're here to help you. Together, we're going to contribute to building a movement."

The Rally

The crowd was many times larger than it had been. I told everyone that this wouldn't be the last Fire Drill Friday and that I hoped they would continue to support and join the student climate strikers who were out every Friday. Then I introduced the woman who had lit the fire and gotten me going, Naomi Klein.

"I was here at the first Fire Drill Friday; there was just a handful of people. And look at you now. And this movement is spreading so fast because this movement is, itself, on fire. And this is something we have to understand. We are not going to win the world that we need for ourselves, for our kids, if we kind of sort of want a Green New Deal. We have to be on fire for it. We have to find that fire inside of us. And we have to get in touch with the life-giving and healing power of fire, right?

BUILDING POWER AND HAVING FUN DOING IT

Annie Leonard

"The reason that this is so exciting and why we don't want to let go of the traction that we're building on Fire Drill Friday is this is the single thing that the climate movement most needs. We know what to do. We have model economic policies. We have innovative green technologies. We have common sense. We have morals. We know how to live in a more environmental and just way. The one thing we're missing is the power to make it so. And that's what we're going to do. That's what we're going to do with the StoptheMoneyPipeline campaign. That's what we're going to do with the Sunrise Movement. And that's what we're going to do with the youth climate strikers and with Fire Drill Fridays. We're going to build the power to make a better future possible. You, all of you watching this, are the one missing ingredient we need to change trajectory.

"If we can work together like this every

single week in our communities all around the country, we'll build the power to change. But another thing will happen, too. It's that we'll have a really good time. And this is important because there's a bit of a narrative out there that being an activist is like being a martyr, and you have to eat cardboard-tasting veggie burgers, and you have to wear sandals year-round, and it's such a hardship life. But the truth is that being an activist, as we all know, feeds our soul. And this is not just romanticism. The data shows this. If you look at the data about what most makes life meaningful, about what most makes you happy once your needs are met. If you don't have a roof and food, more material objects do make you happier. But once your basic needs are met, it's really consistent what most makes you happy. It's a sense of purpose and meaning beyond yourself. It's the act of coming together with others toward shared goals. And it's having lots and lots of friends and social fabric. And the great thing about activism, coming together to fight these big oil companies, coming together to build a green and peaceful future, is that not only do we get to actually save the planet and keep living here, but we get to have way more fun in the process. This last three months has proven that, right? So let's do this!"

"I'm not talking about the fires of annihilation and destruction of the kind that we're seeing in Australia, that we see in California. That is about the misuse of fire. That is about digging up that which should not be burned and putting it in the atmosphere, upsetting the balance of the elements.

"But power is also a healing force, right? In nature, power clears away the debris, the deadwood, and makes room for new growth. That's why indigenous peoples have always used fire to tend to the land, right? So we need to do that. We need to be that fire. And we need to clear away the debris. We're going to clear away the debris of the climate-change deniers funded by fossil fuel money. We are going to clear away the debris of the distracters telling us everything else is more important: Trump's latest tweet, the royal family's fashions, all of that. Clear away that debris. Then we're going to clear away the debris of the doomers, the ones who are telling us it's all too late and we may as well just kind of relax and watch the world burn because they think they're going to be safe. Most of all, we're going to clear away the debris of the dividers. We need to clear away the deadwood and make room for new growth. And we're going to build the most united and powerful movement that we have ever seen.

"The climate strikers in the U.K., they have this slogan. They say, 'Greta was the spark, but we are the wildfire.' And that's what we need to be. We need to be the wildfire."

During our rally, Bill McKibben and a crew of activists had occupied a local Chase bank. They were inside, the police were outside, and we had rigged a connection so that Bill could speak to the crowd.

"I'm inside a Chase bank branch and this really marks, not only the escalation of Fire Drill Fridays, but imparts the real national launch of this Stop the Money Pipeline campaign. We've got to be here because Chase bank is the single biggest funder of fossil fuels on planet earth.

"As the spring goes on, there's going to be hundreds, thousands of demonstrations like the ones that are happening today. And on Monday, we'll be announcing the dates for when we need everybody across America to join in. And it's not just across America.

Bill McKibben speaks to the Fire Drill rally from his occupation inside a Chase bank.

RISE UP: A CALL FROM AN INDIGENOUS LEADER

Ta'Sina Sapa Win Smith

Ta'Sina Sapa Win Smith is an Itazipco, Mnicoujou, and Hunkpapa Lakota of the Cheyenne River Sioux Reservation and a direct descendant of Chief Gall. She is a law student and a frontline environmental and social justice grassroots activist fighting against the Keystone XL Pipeline, the Dakota Access Pipeline, and uranium mining in her ancestral homelands.

"Hello, my relatives. My Lakota name is Black Shawl Woman. I am from the Cheyenne River Sioux Reservation.

"It is no secret that my people endured violent repercussions for standing peacefully and unarmed in 2016 against the Dakota Access Pipeline in Standing Rock. We saw relatives nearly lose eyes and limbs with the use of nonlethal weapons by the state, or, in

Ta'Sina Sapa Win Smith speaks.

my case, being viciously mauled by an attack dog from an illegally permitted private security firm that was on the ground. We've seen the lengths to which the state would go to protect their profit. And right now, again, my ancestral homelands are facing yet another threat, the Keystone XL Pipeline, which will snake right underneath the Cheyenne River for which my tribe is named. Man camps are being constructed as we speak. And I'm not too sure if you all know what man camps are, but they are camps which house the thousands of workers that will be completing the pipeline. One is right next to our reservation border. Now, the thing about man camps is when they're constructed, the rates of missing and murdered indigenous people skyrocket, and drugs become bountiful. And I see this as no coincidence.

"We're living in a time where, across the globe, millions of people are standing up for their right to human survival, the right for clean air, clean water, clean earth, and to save the very ecosystems that are collapsing right now due to the environmental destruction we've created. It is time now that we rise up. Rise up, my brothers and sisters. Rise up and stand against one million species' extinction. Rise up and stand with the original caretakers of these lands, the indigenous people. Rise up and stand against cultural genocide, ecocide, environmental catastrophes, and illegal resource extractions. The time is now that we transition into a new era away from the fuel addiction. The land we stand on has been taken by force to fuel this addiction to pollution, profit, and greed. It is time now, my brothers and sisters, to rise up and stand with the indigenous people. We, the people, are that new era. We are the era that will ensure and protect human survival for the next seven generations. Our house is on fire and we need to put it out."

This morning, our colleagues in the Youth Climate Movement, around the world, led by Greta Thunberg, released a remarkable statement and exactly the same message, demanding that financial institutions begin, not destroying the planet, but investing in the things that will make it work in the future. Everybody is talking from the same book now, and there's no longer any doubt in anyone's mind about what's going on. If banks like Chase bank stopped funding the fossil fuel industry, they could not go on building pipelines. They could not go on making new coal mines and building new LNG terminals and doing all the other things that are making our planet less habitable week by week. So we are so happy to be with you all. The police have just arrived here, but we're going to be here for a while and we hope we get to see you, and, if not, good luck with all that you're doing down there on top of the hill."

It was quite thrilling for us to hear Bill that way.

I then introduced Tara Houska, a tribal attorney, co-founder of the Giniw Collective, former campaigns director of Honor the Earth, and currently engaged in the movement to defund fossil fuels and the years-long struggle against Enbridge's Line 3 pipeline.

"I'm Bear Clan from Couchiching First Nation, and it's great to see all of you," Tara announced. "The last time I was here, speaking behind a rally podium similar to this one, I came from the resistance camp in Standing Rock. I came from seeing my friends teargassed, maced, bitten by dogs. I came from being

Tara Houska speaks.

arrested and put into a dog kennel. I came from watching cops chase Indians on horses across the plains shooting at them. History happening again and again. And this time around, I'm coming from a resistance camp in Minnesota where I've been living two hundred yards off a proposed pipeline route that wants to go through my people's territory, through our sacred wild rice. Line 3 hasn't started bulldozing yet, but they're setting their final permits the next couple weeks. And they're busily arming themselves to the teeth in northern Minnesota, which is really crazy to see. And so my message isn't for the companies that are killing our kids' futures for profit; it's for us. My

BANKS—THE BAD AND THE GOOD

Kat Taylor

"I am so grateful to Jane for this series of stick-in-the-eye protests, and especially this one, for I come to you as a banker of a different stripe. I came to be a banker from a hunch my husband and I had that something was terribly wrong in the banking system. We started Beneficial State Bank to test and pursue that insight, which proved to be true. The banking system is enormously powerful, and it's not serving us or society. It's boring and complicated yet important. It's time to make banking sexy as well.

"We are not separate from each other or from nature, and neither is banking. It's not even our prerogative to reject nature, but hers to reject us. Like watching the worn back pockets of a jilted lover as she walks out of your life, we realize that she will be fine without us and we will in fact have to eat our lunch all by ourselves.

"Banking is also in her back pocket. We

think of banking as the original and most powerful form of crowdfunding. Not that a specific deposit funds a specific loan, but all deposits fund a lending practice that is very impactful. Banking was designed to do that. Within it, we pool our idle cash, call them deposits, and entrust them to the banks. We even add FDIC insurance to make sure that all our deposits are safe and that the banks can collect them readily and cheaply. Our deposits are their rocket fuel. Their lending is our world, and it really matters because the banks have grown to be the biggest sector in our economy. Many of them have well over $2 trillion in assets, and we lend them approximately $12 trillion in our deposits. The banks dwarf every other industry including oil and gas. Moreover, the industry outspends every other industry two to one on political candidacies and lobbying. Of course, we created the banking system in hopes the banks would use our deposits to finance the world in which we actually wish to live. A new economy that is fully inclusive, racially and gender just, and environmentally restorative. Yet we get anything but that from the biggest and most influential banks. As you heard, just since the Paris Agreement in 2016, the leadership of Chase bank alone has financed nearly $200 billion of fossil fuel development. But remember, these are our deposits. We don't have to tolerate the banks using them to destroy our planet. What is fossil fuel without fossil fuel finance?

"Nothing. And therein lies our solution. If we insist the big banks divest from fossil fuel lending, that action will shut down those industries quickly and orderly. We can also move our money to community banks and credit unions, free of implications of fossil fuel disaster, and deprive the bigger banks of their rocket fuel. We can encourage regional banks to bring their lending practice into alignment with our values and get out of not only fossil fuels but payday lending and private prisons. We can insist our regulators take into account the triple bottom line of social justice and environmental well-being alongside financial sustainability. In short, we can change the banking system for good. And we can come 180 degrees from banks burning down the planet to banks building an advanced energy economy. We can move, as William McDonough says, 'not to 0 percent but to 100 percent fabulous.'

"Just as we will take back our government from the corporations who have bought it, we will take back banking because it belongs to all of us, even the unbanked. We will turn the tide on climate change. We will turn gray skies of ash to blue skies and seas. We will take to the promised land of beneficial banking and stop climate change in its tracks."

message is for all of us. We have power. We have so much power. They pull out their tanks when we stand unarmed in front of their machines because they are afraid of us. They use water cannons and attack dogs when we say 'no' with our bodies because they are afraid of us. They kill land defenders in the global south that are unknown, protecting their forests, protecting their territories, because they are afraid of us. We can build the world we want, and we can demand this one changes at the same time. None of this, all around us, operates without us. So pull your money out of the banks like Bill said, run for office, maybe run against whatever office is happening, march in the streets, reduce your carbon footprint, sue the companies, stand in front of the machines. More of us need to stand in front of the machines. And also, when you're doing all of those things and finding how you can contribute, reconnect to the earth, deeply, irrevocably, reconnect to the earth. Don't just retweet or share this selfie; that's not good enough. Love the air; the air loves you. Love the sun; the sun loves you. Love the water; the water loves you. The answers are there; we just have to listen. The answers to this crisis are right in front of us. It is a life in balance, not a life of greed. It's a life of love, not of lying to ourselves, as we traumatized animals in neat little packages, click on cell phones that are made by unpaid hands somewhere else that we can't see. We have to have a life of stewardship, not just have endless consumption. Stuff will never fill our hearts. Stuff will never make us feel complete. Do not romanticize my words either. Native peoples, we've figured some things out. And all of you used to know that; we just have to remember it together. We all have the answers, so remember who you are, don't fall into complacency or apathy, you're out here doing something, that's so beautiful to see. Build communities; stand with love. Stand brave, stand fearless, because we're in this together."

When Kat Taylor, co-founder and co-CEO of Beneficial State Bank, took the stage, she, too, urged us to stand together to pressure big banks to divest from fossil fuel lending. And then, suddenly, Kat burst into an anti-pipeline song sung to the tune of "I Heard It Through the Grapevine." Not what you'd expect from a banker!

Kat Taylor speaks—and sings—to the crowd.

June Diane Raphael, making a return visit to Fire Drill Fridays with her five-year-old son, Gus, introduced Eriel Tchekwie Deranger.

After telling the crowd about the Teck

AN IRISH STORY

Martin Sheen

"The Irish tell the story of a man who arrives at the gates of heaven and asks to be let in. Saint Peter says, 'Of course, just show us your scars.' The man says, 'I have no scars.' Saint Peter says, 'What a pity. Was there nothing worth fighting for?'

Martin Sheen addresses the crowd. Susan Sarandon sits behind him.

We are called to find something in our lives worth fighting for. Something that unites the will of the spirit with the work of the flesh."

And then Martin began to paraphrase an excerpt of a famous poem by the Bengali poet Rabindranath Tagore called "Gitanjali 35." "Something that can help us lift up this nation and all its people to that place where the heart is without fear and the head is held high, where knowledge is free, where the world has not been broken up into fragments by narrow, domestic wolves, where words come out from the depths of truth, and tireless striving stretches its arms towards perfect, where the clear stream of reason has not lost its way into the dreary desert sands of dead habit, where the mind is led forward by thee into ever-widening thought and action. Into that heaven of freedom, dear Father, let our country awake."

BECOMING SOCIAL RISK

Rebecca Adamson

Rebecca Adamson is an American Cherokee businessperson and advocate. She is the former director and president of First Nations Development Institute and the founder of First Peoples Worldwide.

"I've been an indigenous economist my whole life, working around getting principles of sustainability, balance, and harmony into this crazy, corrupt economic system. I'm going to give you a couple of statistics: Today, as we stand here, 39 percent of all oil, gas, and mining is on indigenous territories. Forty-six percent of the future reserves

Rebecca Adamson speaks.

are on indigenous territories. About four indigenous leaders a week are killed for defending their lands, defending indigenous rights, defending the environment, and it's not a co-incidence.

"We need to come together and look at the market. I want to talk about the effect you're having on the financial market. I've been a social investor for about thirty years now. That means when I go to make a decision on my investments, I look at the environmental performance, I look at the social performance before I make my investment decisions. Social investing has turned into ESG [environmental, social, and governance] investing, which is growing exponentially in the market. You, here today, are protesters, engaged in civil disobedience, any number of things, but in the market you are recognized as social risk.

"I was asked by the Standing Rock Tribe to help lead the divestment strategy for getting people and banks out of investments in DAPL [Dakota Access Pipeline]. And within the first two weeks of work against DAPL, the tribe had over one million hits on their website. For our day of solidarity, we had three hundred events across fifty states and in Paris, London, and Australia. Everywhere around the world, we were having solidarity events. We had seventeen banks that were funding DAPL. They lost $4.4 billion in account closures. That's significant to these guys. You're now in the market, and you're now social risks that could be quantified. We had three banks pull out and pull their loans. The stock of ETP [Energy Transfer Partners] dropped 60 percent in value.

"A 60 percent drop in your value is called material risk. ETP lost $7.5 billion at the DAPL site. Seven point five billion dollars is material risk. It's when the SEC, Securities and Exchange Commission, wants you to make a report on the risk to your investors. It's when the cost of capital starts to go up and your money starts to cost you more to destroy Mother Earth.

"Social risk, in the broadest sense of your protests, of your campaigns when you shut down or protest against a site, is between $20 and $30 million a week of social losses to the market. Twenty to thirty million dollars a week. There's $25 billion in tied-up or stopped mining operations as we talk. Those are lost assets. That's material risk, all brought about by social risk.

"Seventy-three percent of the risk that corporations face today is 'nontechnical,' which means you all. Nontechnical risk is community protest. It's letter writing. It's campaigns; it's stopping their work, tying it up. Right now, nontechnical risk is at seven; it's the second-highest risk in the financial market.

"So protest. Be social risk."

Eriel Tchekwie Deranger speaks.

Resources tar sands mine that threatened her territory, Eriel ended by saying, "Join our struggle as we fight the pipelines. We are the strong. We are the resilient. We are the woke. We are the indigenous. Our bodies and our spirits warmed by the land. Our minds nourished by the living earth. We are the reimagination of today. We are the visionaries of tomorrow. We are the keepers of the ancestors. We are the woke."

Reading this now, knowing that her peoples and their supporters had, in fact, achieved the unimaginable and defeated the immense mine, makes her words all the more stirring.

After the rally, we marched toward the Capitol, where the civil disobedience was going to take place.

For the first time, there was a solid crowd between

The group heads to engage in civil disobedience on the Capitol steps. Susan Sarandon, Joaquin Phoenix, Omékongo Dibinga, Jane, Eriel Tchekwie Deranger, and Naomi Klein.

the rally and the steps, and as I looked up, I saw an enormous number of people already on the steps and thought, **Oh, Lord, clearly another group of protesters have planned on occupying the steps and there won't be room for us.** I didn't understand what was going on until Ira explained to me that everyone in the crowd and on the steps was part of Fire Drill Fridays. They had just left the rally early to get there. It turned that out well over three hundred

The crowd moves to watch the speakers, introducers, and activists risk arrest (note the crowd size compared with week one).

people were risking arrest that day. I was stunned. What an amazing way to end this leg of the Fire Drill Fridays journey.

I waited to see people start to get arrested, and then a large group of us, including Susan Sarandon, who had come down that morning from New York to join us, marched about three-quarters of a mile to the Chase bank where Bill McKibben and about a dozen activists were sitting in with a huge "Stop the Money Pipeline" banner. Standing outside the large bank windows, we waved and chanted our support. A dozen or so police officers were also gathered outside the bank, as it turned out, waiting for us to leave before arresting everyone inside. There was press there as well, and the occupation of Chase got a lot of coverage.

After an hour, we marched back to the Capitol, where, because there were so many who wanted to be arrested, the Capitol Police had decided to process them next to the steps.

The group stands in front of a Chase bank demanding an end to fossil fuel funding in conjunction with Jane Fonda's Fire Drill Friday rally. JPMorgan Chase is one of the largest sources of capital to the fossil fuel industry in its quest to drill oceans, frack our land, and build more pipelines. The group was supporting allies who were inside the bank, risking arrest. The group includes Susan Sarandon, Vanessa Vadim, Jane, and Annie Leonard.

Jane hugs Joaquin as he is released, on-site. He was the last one to be released.

It was hours before everyone was released. Naomi, Joaquin Phoenix, and Martin Sheen were among the last, and I was there to hug and thank them and about a hundred other new friends I'd made during the fourteen Fire Drill Fridays, people like David Fenton, Gene Karpinski, Kallan Benson, Sebastian Medina-Tayac, and Erika Berg, along with the ones who had run the digital operation, Firas and Carla, and our logistics helper, Sam. Jerome Foster gave me a Christmas ornament that said, "I ♥ My Godmother," which touched me a lot.

Then it was back to the hotel to finish packing. I

would return to California with Debi early the next morning to begin preparing for the next phase of Fire Drill Fridays . . . oh yes, and **Grace and Frankie.**

What Can I Do?

As Bill McKibben has written, "Money is the oxygen on which the fire of global warming burns." We've got to move our money **out** of fossil fuels and deforestation and **into** climate solutions. Luckily, "moving your money" has never been easier. There are a number of ways that you may be connected with the financial system: a bank account, a credit card, your insurance company, or your retirement account or investment portfolio. Let's walk through each of these areas and talk about how you can make sure your money is fossil-free.

First up is your bank account. Big banks like JPMorgan Chase are some of the largest funders of fossil fuels, deforestation, and general climate destruction. The best way to find out if your bank is fueling the climate crisis is by searching for the latest **Banking on Climate Change** report by the Rainforest Action Network, which every year updates its list of banks that are doing the most to wreck our climate. In 2019, JPMorgan Chase was at the top of the list, followed by Wells Fargo, Citi, Bank of America, RBC, and Barclays.

What's the alternative to a climate-wrecking

megabank? The DivestInvest guide to banking (divestinvestguide.org) will guide you through the steps of picking a new bank, closing your account, and moving your money. A useful resource for finding a new bank is Green America's "Get a Better Bank" tool (greenamerica.org/getabetterbank). Often, the best option in your area will be a local credit union. You could also check out a new, fossil-free online bank, like Aspiration, which reimburses you for ATM fees nationwide.

Once you've cleaned up your bank account, let's take a look at those credit cards. Surprise! Just when you thought you'd left your megabank for good, you look in your wallet and realize that your credit card is connecting you right back to the bank you wanted to get away from in the first place!

Let's face it: Even if you pay your credit card bill on time every month, every time you swipe your card, a small percentage of your purchase is going back to a dirty bank. You're also lending that bank legitimacy every time you pull out that credit card (JPMorgan Chase isn't going to take our consumer complaints very seriously if we all hang on to those Sapphire cards). The good news is, there are all sorts of ethical credit cards out there; you can even find a card that directs a percentage of every purchase to supporting environmental groups, restoring ecosystems, or planting trees. Green America has a list of ethical options that can get you started (greenamerica .org/responsible-credit-cards). If you end up switching your card, make sure to tell your old bank that you're doing it because they're financing fossil fuels. Better yet, make a video of you cutting up your card, post it online, and tag the bank you're leaving. That will get their attention!

Next up, insurance companies. You can find out if your insurance company is financing fossil fuels, and if so, find an alternative, at websites like Insure Our Future (insureourfuture.us) and Unfriend Coal (unfriendcoal.com/insurance).

Now let's tackle those retirement accounts and investments. The best place to start is the DivestInvest website (divestinvest.org). They've got a bunch of useful links to walk you through the steps of greening your portfolio. Want to check out a specific ETF or mutual fund? Check out Fossil Free Funds (fossilfreefunds.org). Their search function allows you to look at specific funds or asset managers, like Vanguard, Fidelity, and State Street. They've also got a list of the cleanest funds on the market and clearly list their cost, expected returns, and all the info you need to make a wise investment.

Here's the bonus: Research shows that getting out of fossil fuels is likely to increase your returns. According to the financial research firm MSCI, between 2010 and 2017 fossil-free portfolios **outperformed** portfolios that hung on to coal, oil, and gas. As the Institute for Energy Economics and Financial Analysis put it in a recent report, "The fossil fuel sector is shrinking financially, and the rationale for investing in it is untenable." Turns out going fossil-free is good for both the planet and the pocketbook.

Finally, we all know that individual action isn't enough. Persuading your alma mater, religious community, pension fund, city, state, or another institution you are associated with to divest is one of the most powerful ways we can start shifting our economy in the right direction.

I'll leave you with the words of one of my great heroes, Archbishop Desmond Tutu, who won a Nobel Peace Prize for his work to end apartheid in South Africa and has been a proud supporter of the fossil fuel divestment campaign: "Just as we argued in the 1980s that those who conducted business with apartheid South Africa were aiding and abetting an immoral system, we can say that nobody should profit from the rising temperatures, seas and human suffering caused by the burning of fossil fuels."

Amen, Archbishop. Ending the financing of fossil fuels and deforestation is a key solution to the climate crisis. Moving your money is a great place to start.

Jane hugs Annie Leonard as she speaks about the future of Fire Drill Fridays.

CHAPTER SEVENTEEN

Fire Drill Fridays:
Going Forward

By Annie Leonard

Early on in Fire Drill Fridays, Jane asked Senator Ed Markey how concerned citizens can advance climate solutions. He told her to build an army and make it big. We agree!

The climate movement already has **almost** everything it needs to transition to a renewable energy economy that will make our country healthier, safer, fairer to all, and more prosperous. We have model regulatory policies, innovative green technologies, economic rationale, and a strong scientific and moral imperative. The one thing we lack is the power to drive the changes we need. That's starting to change, and you, dear reader, can be a part of this! **We need you**. We know that the power of millions of people engaging in an issue will create change, because we have seen it happen before.

In 1970, the United States celebrated its first Earth Day, ushering in a decade of unprecedented attention to and solutions for environmental problems. Do you know how many people participated in the first Earth Day in the United States? **Twenty million!** Twenty million people marched in the streets, built community gardens, cleaned up beaches—did visible actions to demand environmental protection. In the decade that saw the first Earth Day, the federal government created the Environmental Protection Agency, or EPA (even under a Republican president), and passed the most ambitious package of laws to protect the environment ever, including the Clean Air Act, the Clean Water Act, the Endangered Species Act, the Safe Drinking Water Act, and the Toxic Substances Control Act.

The government created the EPA and passed all these laws not because it was moved by the scientific arguments but because 20 million people demanded it. We urgently need that same level of ambitious action now on climate, and the way to get it is millions joining together once again. The good news is that it's already starting. The youth-led 2019 climate strikes were the biggest climate protests to date, with estimates of more than 7.5 million people participating all around the world. That's a great start—and this year, let's do even more!

FIRE DRILL FRIDAYS CALLS ON PEOPLE TO SPEAK, VOTE, AND ACT—AND WE ARE HERE TO HELP!

SPEAK: It's time to turn the volume up on conversations about climate until we are too loud for political and business leaders to ignore. We're in this odd situation in our country in which more people than ever are worried about climate change and want government action, but we still don't talk about it much. I get it; talking about the climate emergency can be intimidating and overwhelming and can certainly dampen the mood at a social gathering. **But we have to talk about it to elevate it on the public agenda**. So share what you've learned in this book. Write letters to the editor and letters to your elected officials. Start a climate emergency team at work, at school, at church, or in your neighborhood. Chat with people in the grocery store aisle, with your family and neighbors. Invite friends over to watch a Fire Drill Friday teach-in and then talk about ways to get involved.

There are lots of resources to help you gain comfort when talking to your neighbors or to your elected officials about climate change. Learn how to talk to climate skeptics, at Skeptical Science (skepticalscience.com), which has great responses to the myths and misunderstandings about climate change. Living Room Conversations (livingroomconversations.org) has guidance for how to talk to people with very different opinions from yours, helping you to find common ground despite your contrasting politics.

Our collective hesitation to talk about the crisis leaves the culprits free to go about their destruction unchallenged. So start by speaking up and speaking out!

VOTE: Seriously, **vote for climate leaders!**

Don't know where your elected representatives stand on climate issues? There are lots of resources to help you find out about the voting records of elected officials and candidates. First, check to see if they have signed the No Fossil Fuel Money Pledge at NoFossilFuelMoney.org. Vote Smart (votesmart.org) lets you type in your address and instantly find all your federal and state representatives and their voting records, issue positions, and campaign finance information.

But don't stop at the state and federal levels; research and vote down ballot too! There are many elected positions where real climate leaders can make a difference: City councils, boards of supervisors, district attorneys, state legislators, zoning boards, and judges all are in positions where they can help stop fossil fuels and aid in the transition to a clean energy economy.

Many national organizations provide information on candidates' and elected officials' environmental policy track record. The League of Conservation Voters tracks environmental voting records, and the Sunrise Movement tracks which candidates support a

Green New Deal. NoFossilFuelMoney.org identifies candidates who have refused donations from fossil fuel interests, so they should not be beholden to them once in office.

These are great resources, but generally they don't include all the candidates for local offices, so you have to do some research yourself. To help elect true climate leaders, divide up the candidates' names among a group of friends so each can research different ones. You may have to call candidates' offices to find out their positions on issues important to you like a Green New Deal, a just transition off fossil fuels, ending public subsidies for fossil fuels, holding fossil fuel corporations accountable, and respecting indigenous rights. Once you and your friends have evaluated the candidates, please share your findings to help others identify climate leaders and climate laggards.

Focusing on local and state elections allows us to keep building momentum even when passing climate legislation at the federal level is stuck. All over the country people in cities and states are organizing to elect climate leaders on city councils, zoning boards, and port commissions and to get initiatives on the ballot that ban fracking, limit new oil and gas expansion, incentivize renewables, and support state-level Green New Deals.

Some see voting as insignificant, but it really does matter. That's why some opponents of change try to prevent you from voting by shutting polling stations, requiring specific voter identification, and disenfranchising former felons who have served their time. Between elections, support campaigns to make it easier to vote. Programs like automatic voter registration, vote by mail, and mandatory time off on Election Day can help get more people to the polls. The more people who vote, the more likely that successful candidates will reflect the public interest. So please register, research, vote, and recruit friends to do so too.

ACT: This book is full of ideas, big and small, for ways each of us can help drive positive change. Scan the ideas in the "What Can I Do?" sections to find those that best match your skills and interests. There are so many ways to get involved that you really can choose one just right for you. No need to do this alone. Working for change is fun and sustainable and has a bigger impact when we do it with others. Connect with an organization that brings you into a community of people who share your concerns. If you don't find a fit with one of the many organizations we've mentioned in this book, recruit some friends and start your own.

ACT: Hold your own Fire Drill Friday.

Join or host a Fire Drill Friday event in your community: To make this movement grow, we need people across the country ready to join us! The Fire Drill Fridays team provides support and guidance for you to take effective and meaningful action from wherever you are, from hosting your own Fire Drill Friday event to helping support the actions happening nationwide. We need you to win this. Sign up at FireDrillFridays.com.

Join Jane and Greenpeace: We're going to continue to hold Fire Drills around the country. Sign up at FireDrillFridays.com to find out where we'll be, and if we are near you, please join us in person!

We know this book is coming out at a time of great uncertainty in our country. We don't know who the next president will be, and the choices, as of the time of writing, run from one with decades of pushing real climate solutions including supporting a Green New Deal and an end to fossil fuels to one who maintains climate change is a hoax and is ushering in a free-for-all for fossil fuel companies eager to plunder the planet. But we do know one thing: Regardless of who is in that office, people power is needed to make him act. Whoever is in the Oval Office in 2021 will face a country with paralyzing polarization and pressing needs. The issues that get addressed first and best are those that have robust public movements demanding action. It is our job to ensure that the president's priority agenda includes immediate and ambitious action to address the climate emergency.

This is our moment. For the first time in our country, a majority of people are with us in sharing our concern and grief about climate and our desire for government action. It's time now to put that concern into action and solve this crisis. Every day we delay, the challenge ahead of us grows. So let's do this.

Acknowledgments

Like the Fire Drill Fridays themselves, this book is truly a collective endeavor, and I want to acknowledge and thank all the people who have played a role in getting it done.

First of all, there is Annie Leonard. She was central to making Fire Drill Fridays so effective, and this book would not even exist, the information you receive here wouldn't be as thorough and diverse, had she not made certain we had the speakers we did. She also was in charge of gathering the information for the "What Can I Do?" sections of the book, and she wrote the appendixes, "An Introduction to Understanding the Climate Emergency" and "Civil Disobedience." She is a true activist leader who lives her values and always exhibits wisdom, compassion, and generosity. I am so very grateful for all she brought to this project.

This book would not exist were it not for the invaluable wisdom and information about the climate crisis that the speakers at the Fire Drill Fridays and Teach-Ins shared. I am deeply indebted to them all: Rebecca Adamson, Lindsey Allen, Yvette Arellano, the Reverend William Barber, Maude Barlow, Phyllis

Bennis, Kallan Benson, Diamonté Brown, Alfred Brownell, Liz Butler, Rachel Carmona, Imam Saffet Catovic, Jasilyn Charger, Keya Chatterjee, Donna Chavis, Veronica Coptis, Whitney Crowder, Eriel Tchekwie Deranger, Omékongo Dibinga, Ellen Dorsey, Asali DeVan Ecclesiastes, Catherine Flowers, Jerome Foster II, Josh Fox, Cherri Foytlin, Jim Goodman, Mary Grant, Michael Leon Guerrero, Roshi Joan Halifax, the Reverend Fletcher Harper, Katharine Hayhoe, Hana Heineken, Von Hernandez, John Hocevar, Tara Houska, Dolores Huerta, Jennifer Jacquet, Charlie Jiang, Khadija Khokhar, Naomi Klein, Winona LaDuke, Abigail Leedy, Annie Leonard, Jessica Loya, Gaurav Madan, Mark Magaña, Bill McKibben, Jansikwe Medina-Tayac, Rolando Navarro, Matt Nelson, Tamara Toles O'Laughlin, Alice Brown Otter, Denise Patel, Sunni Patterson, Ai-jen Poo, Monica Ramirez, Katie Redford, Janet Redman, Garett Reppenhagen, the Reverend Malik Saafir, Ricardo Salvador, Sarah Schumann, Laura Turner Seydel, Rabbi David Shneyer, Rolf Skar, Samantha Smith, Clint Sobratti, Saket Soni, Sandra Steingraber, Kerene Tayloe, Gail Taylor, Kat Taylor, Kristin Taylor, Anabell Castro Thompson, Heather Toney, Krystal Two Bulls, Joe Uehlein, and Vanessa Vadim.

My editor at Penguin Random House, Ann Godoff. Thanks for supporting this book from the get-go.

Danelle Morton, thank you for helping me with the editing, dealing with the transcripts, and, most of all, coming up with the question that you felt should be the title: "What can I do?"

Linda Loewenthal, my agent, who fought for the book all the way.

Those who would prefer we not know the truth about the climate crisis never waste an opportunity to disparage celebrities when they use their platforms to raise awareness. They do this because they know that when famous people speak out, the message reaches a broader audience. I view celebrities like repeaters, the tall antennas you see on mountaintops. Repeaters are used to extend the signals in the valley so that they can cover longer distances. So I want to thank my "repeater" friends who helped ensure that the experts, the voices from the front lines, were heard far and wide. Of course, some of the "repeaters" are experts in their own right, like Gloria Steinem, Eve Ensler, Bobby Kennedy, and Ted Danson. My deep thanks to them and to Iain Armitage, Rosanna Arquette, Ben Cohen, Abigail Disney, Sally Field, Brooklyn Decker, Jerry Greenfield, Marg Helgenberger, Manny Jacinto, Catherine Keener, Diane Lane, Piper Perabo, Joaquin Phoenix, June Diane Raphael, Susan Sarandon, Paul Scheer, Taylor Schilling, Kyra Sedgwick, Martin Sheen, Maura Tierney, Lily Tomlin, and Amber Valletta. And I'm grateful to Emma's Revolution and Sweet Honey in the Rock for bringing their music and heart to the actions.

There were people who contributed the issue briefs we received to prepare for the teach-ins and rallies and this book. They proofread and added expertise when called upon and I am very grateful: Ira Arlook, Maude Barlow, Phyllis Bennis, Daniel Brindis, Madeline Carretero, Mike Cavanaugh, Megan Davis, Niaz Dorry, Ellen Dorsey, Jodie Evans, David Fenton, Peter Gleick, Julie Gorecki, Mary Grant, Jamie Henn, John Hocevar, Ayana Johnson, Melinda Kramer, Steve Kretzmann, Annie Leonard, Monique Mikhail, Karen Nussbaum, Melina Packer, Katie Redford, Collin Rees, Sandra Steingraber, Sam Waterston, Ta'Sina Sapa Win Smith, Emira Woods, Janene Yazzie, and Joanna Zhu. Tim Aubry took all the photographs at the Fire Drill Fridays that are in this book.

My thanks to all those who were at the launch meeting where the Fire Drill Fridays plan was hatched. Every one of you contributed to its success: Ira Arlook, Madeline Carretero, Keya Chatterjee, Charlie Jiang, Annie Leonard, Jose Martinez-Diaz, Sam Miller, Noor Mir, Karen Nussbaum, Erich Pica, Janet Redman, Deirdre Shelly, and Reverend Yearwood.

Then there is the core team for Fire Drill Fridays. I am indebted to them because, although all of them didn't participate directly in the development of this book, the work they did for Fire Drill Fridays is what made Penguin Random House want to publish it.

Maddy Carretero is the person at Greenpeace who develops associations with celebrities, influencers, and funders, involves them in the organization's actions when appropriate, and organizes the events with them. For many months, Maddy was my primary go-to Greenpeace person for what needed to be said at every rally and whom we could invite to join.

It's possible that we could have done all those Fire Drills every Friday and few people would have known about it. Ira Arlook made sure the world knew about it. We were joined at the hip for at least two months, and I am deeply grateful for what he did for us.

Cultivate Strategies' Carla Aronsohn, with Firas Nasr, created our website in about ten days, constantly uploading and refreshing the content and keeping track of our numbers. Firas designed the website graphics and did all the livestreaming of the rallies, often leading our chanting and preparing us through meditation and shouts for the action.

Samantha "Sam" Miller, with DC Action Lab, was the lead logistics person—getting the permits, setting up the stage and sound system, dealing with the police, handling speakers' transportation, lodging, and jail support—she basically ran the weekly physical operation with rarely a glitch. Maria Brescia-Weiler was very helpful in assisting with all aspects of the logistics. Robby Diesu, also with DC Action Lab, joined the team for the final series of Fire Drill Fridays.

Karen Nussbaum was our go-to adviser when it came to finding ways to bring labor into our climate actions. Not to mention the moral support she brought me by always being there.

Jose Martinez-Diaz, program engagement director

at Greenpeace in charge of its large campaigns, worked with Fire Drill Fridays to ensure we were in sync with the various Greenpeace projects and helped plan and facilitate the vision for our national rollout. A solid, wise adviser.

The artist Vy Vu created original posters for each Fire Drill Friday, working in the evenings after school, and I am so thankful for her talent and focus.

I don't know what I would have done without Debi Karolewski, who came with me from California, made sure I had everything I needed on every level, and was by my side for every action, preparing my notes and making sure they were in my hands at the right time and out of them for the arrests.

Mary "Lulu" Williams quickly became part of the team, first as a marshal during the actions, then as one of the set-up-and-take-down staffers.

I want to thank the thoughtful and vigilant Mara Verheyden-Hilliard, my lawyer during my time in D.C.

I owe a special thanks to the Greenpeace USA office for letting us take over their conference room for four months and the Greenpeace Climate Team for the climate briefings.

Appendix A:
An Introduction to Understanding
the Climate Emergency

I asked Annie Leonard to write this chapter, which I feel helps us understand the science of the crisis and the urgent solutions.

At one level, the causes and solutions to climate change are complex, but at another level they are pretty simple: We need to stop the problem and advance solutions. And we need to do both ambitiously and aggressively because we don't have a lot of time.

At the heart of the problem are emissions of carbon dioxide (CO_2) that build up in the earth's atmosphere, trapping heat and slowly increasing global temperatures. These CO_2 emissions mainly come from burning fossil fuels: coal, oil, and gas. Other things, like changes in land-use patterns, deforestation, and methane releases from melting permafrost in the Arctic and the infamous cow burps, also contribute to climate change, but the biggest driver by far is CO_2 emissions caused by burning fossil fuels.

Scientists constantly track levels of CO_2 in the

CO_2 Emissions from Developed Fossil Fuel Reserves, Compared to Carbon Budgets (as of Jan. 2018) Within Range of the Paris Goals

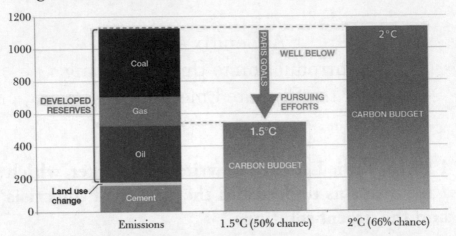

Source: Oil Change International (OCI) 2019; analysis of data from Rystad Energy, International Energy Agency, World Energy Council, and Intergovernmental Panel on Climate Change. Kelly Trout and Lorne Stockman, "Drilling Towards Disaster: Why U.S. oil & gas expansion is incompatible with climate limits," Oil Change International, January 2019, http://priceofoil .org/2019/01/16/report-drilling-towards-disaster.

atmosphere and can figure out CO_2 levels from the distant past through methods like sampling air bubbles trapped in glacial ice. They measure concentration of CO_2 in parts per million, abbreviated as ppm. Before the Industrial Revolution, in the mid-eighteenth century, global atmospheric concentrations of CO_2 hovered around 280 ppm. Since the Industrial Revolution, CO_2 levels have increased steadily, recently reaching 415 ppm. We simply don't know what

Top Countries by Increase in Oil and Gas Production to
2030 (over 2017 Baseline)

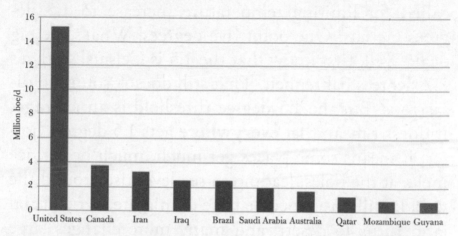

Source: OCI 2019; using data from Rystad Energy. Kelly
Trout and Lorne Stockman, "Drilling Towards Disaster:
Why U.S. oil & gas expansion is incompatible with climate
limits," Oil Change International, January 2019,
http://priceofoil.org/2019/01/16/report-drilling-towards
-disaster.

a 415 ppm world holds for us, because we, as a species,
have no experience living under this condition.

Scientists have identified CO_2 levels of 350 ppm
as a threshold for maintaining a stable planet for
human society to thrive (hence, the climate organization called 350.org, co-founded by Bill McKibben).
So that's our target: 350 ppm. But we're going in the
wrong direction, with CO_2 emissions still increasing.

Climate scientists often compare current global

temperatures with average temperatures before the Industrial Revolution. Globally, scientists have called for limiting temperature increase to 1.5 degrees Celsius. One point five degrees! What's the big deal? Well, first know that the 1.5 is Celsius, so that's 2.7 degrees Fahrenheit. That still doesn't sound so disastrous. But the 1.5-degree threshold is an **average.** It doesn't mean that **everywhere** gets 1.5 degrees hotter; it means some places get much, much hotter, ice melts at the poles, leading to sea level rise around the world, millions of people have to migrate, agricultural land turns to desert, and many more changes large and small to the planet as we know it.

The global temperature increase so far is only about 1 degree Celsius. Already that brought us devastating increases in extreme weather, fires, droughts, species die-offs, disruption in animal behavior, and all sorts of changes to how our planet functions.

Tragically, many of the countries and people most impacted by these changes are also those that contributed the least to putting that CO_2 into the atmosphere. For this reason, those of us in heavily industrialized countries like the United States that materially benefited from all that fossil-fuel-fueled development have an extra responsibility to address the problem and help those most impacted—because we can.

So this is serious. The science is clear. We know what the problem is, and we know what we need to do. And the good news is that change is possible!

In 2018 the Intergovernmental Panel on Climate

Change, the world's top climate scientists, issued a report laying out what's needed to avoid the worst climate impacts. In summary, it said that we have to reduce human-caused emissions of CO_2 by about half from 2010 levels by 2030 and then keep reducing down to zero by 2050. Because burning fossil fuels is the biggest contributor to human-caused CO_2 emissions, what the IPCC study means in practice is that we have to reduce fossil fuels by about 50 percent by 2030 and then keep reducing.

We Already Have More Fossil Fuels Than We Can Safely Burn

That's why Fire Drill Fridays is calling for an immediate halt to **new** fossil fuel development. Every new fossil fuel project, permit, pump, and pipeline makes the challenge ahead of us harder. And it is already going to be very, very hard. Step number one must be to stop making the problem worse to buy us some time to make it better.

The United States Is the Epicenter of Global Fossil Fuel Industry Expansion

Stopping new fossil fuel emissions is key, but we also need to invest in big, bold solutions, increasing the availability of clean, safe renewable alternative energy, as well as improving energy efficiency so we use less energy overall. We need to invest in large-scale infrastructure for public transportation, electric vehicles, and climate-resilient communities. And to protect the oceans and forests and adopt regenerative agricultural practices—all of which will help mitigate climate change.

That's why Fire Drill Fridays is also demanding a Green New Deal, which calls for a massive, unprecedented, society-wide effort to power our economy with clean renewable energy and transform other sectors of society to be more sustainable. The ending of fossil fuels and passing a Green New Deal is a winning combination. And if we do both of these right, we can avoid climate disaster while also remaking our economy to be fairer, healthier, and more secure for everyone. This is the ultimate win-win combination, with prosperity and well-being for all.

The good news is that there are many solutions for clean renewable energy sources that are growing fast. There's also massive potential for energy efficiency and conservation through better building design, more efficient and accessible public transportation, and changes to how we use land, grow food, and make all our stuff. All of these can create good union jobs that sustain families and the planet. Experts and activists know what we need to do; the obstacles to ambitiously addressing today's climate emergency are not technical but political. That's why we created Fire Drill Fridays.

Some critics point out that the changes needed to address climate change are expensive. That's true. But inaction has a far bigger price tag; it's expensive to keep rebuilding after disasters, try to protect coastal cities from rising sea levels, manage mass climate migration, and respond to public health impacts from fossil fuel pollution and climate change. The U.S. Environmental Protection Agency estimates that by the end of this century the impacts of climate change will cost our economy hundreds of billions of dollars annually. It's deadly, too. At least a quarter of a million people already die per year due to climate change, and that number may increase to half a million per year by 2030, according to estimates from institutions like the World Health Organization and Global Humanitarian Forum.With all that we know today, and all the resources our wealthy country has, delaying action on climate because of bogus worries about cost are at best disingenuous and at worst downright immoral.

Fire Drill Fridays is working to force our leaders to lead. Together we are demanding action to keep our communities safe and our planet functioning. We are educating ourselves, organizing with friends and allies, and working together to elect candidates willing to replace those blocking progress. We don't have a moment to lose, and we invite you to join us.

Appendix B:
Civil Disobedience

Fire Drill Fridays were my first (conscious) experience with civil disobedience. I have been arrested for distributing copies of the Uniform Code of Military Justice to soldiers on military bases, but if that was considered civil disobedience, I wasn't aware of it. So while I understand the power that civil disobedience holds, I asked Annie to write a chapter about it. If we're saying that civil disobedience needs to become the new norm, it's important for potential Fire Drillers to fully understand what that means.

Civil disobedience embodies the best of what America can be. The 1773 Boston Tea Party was a form of civil disobedience, and in the roughly 250 years since, civil disobedience has been used across the United States by labor unions, women's groups, immigrant rights groups, faith leaders, indigenous activists, Black Lives Matter activists, and generations of students to raise attention to an injustice, build public support for a cause, and pressure decision makers to act.

Many of us grew up hearing stories of courageous people who engaged in civil disobedience to advance moral and just causes. Martin Luther King Jr. engaged in peaceful civil disobedience to advance voting rights, civil rights, and racial equity. Rosa Parks famously sat down on a bus and refused to relinquish her seat to a white passenger to protest entrenched racist segregation laws. Dolores Huerta and Cesar Chavez organized nonviolent strikes, marches, boycotts, and fasts to advance farmworkers' rights.

Greenpeace, the organization I lead in the United States, started with an act of civil disobedience. In 1971, our founders rented an old fishing boat and set off from Vancouver, Canada, aiming to prevent a bomb detonation. Since then, we have engaged in civil disobedience to protect sea life, safeguard forests, stop oil drilling, ban hazardous chemicals, protest war, advance climate solutions, and many other good causes.

In its most fundamental form, civil disobedience is breaking a law that itself is seen as an unjust law—like withholding taxes that pay for war or sitting at a racially segregated lunch counter. But civil disobedience can also take the form of refusing to follow police orders as when Fire Drill Fridays participants refused to vacate the steps of the U.S. Capitol or a public road in defiance of police instructions. Civil disobedience can include physically putting one's body between someone causing harm and the harm itself, such as when Greenpeace sailed into a nuclear test zone, or when the lone man stood in front of a tank in Tiananmen Square.

In some cases, civil disobedience involves engaging in a prohibited act, such as the humanitarian volunteers who have been prosecuted in relation to their leaving water for migrants on the treacherous journey across the desert along the U.S.-Mexico border. It can be creative, like the Extinction Rebellion activists in Portland, Oregon, who repeatedly build beautiful gardens on the train tracks that carry Alberta tar sands oil through their state. It can be pretty straightforward, such as the millions of students who strike from school every Friday. And it can be highly skilled and require extensive training, like when Greenpeace activists hang off bridges for days at a time to block oil-related equipment from passing through.

In launching Fire Drill Fridays, we felt compelled to bring the power of civil disobedience to address the climate emergency. For more than three decades, climate scientists and activists have conducted research, lobbied elected officials, gathered signatures on petitions, organized marches, written books, made documentaries, created educational programs, run campaigns, lectured, pleaded, and divested. And in those three decades of skilled and passionate work, carbon dioxide emissions have increased, global temperatures have risen, and no elected leader has taken action even close to what the science demands. None of our advocacy-as-usual is delivering results at the level and speed needed. And we're running out of time.

That's why this is an emergency. And that's why we believe it is time for widespread civil disobedience to force action.

You may expect someone from Greenpeace to say this, but I assure you we are not alone. It is a new day in the climate movement, and a growing number of people are ready for civil disobedience.

In 2019, over fifteen hundred scientists from more than twenty countries called for civil disobedience, explaining, "We believe that the continued governmental inaction over the climate and ecological crisis now justifies peaceful and nonviolent protest and direct action, even if this goes beyond the bounds of the current law." And as we were organizing regular civil disobedience with Fire Drill Fridays, Christiana Figueres, the former executive secretary of the UN Framework Convention on Climate Change—in other words, not a radical activist by any means—announced that "it's time to participate in non-violent political movements wherever possible."

In an inspiring new book on climate, **The Future We Choose**, co-written with Tom Rivett-Carnac, she said, "Civil disobedience is not only a moral choice, it is also the most powerful way of shaping world politics."

Millions of young people, and increasingly adults, too, have heeded the call of the teenager Greta Thunberg, who urged us to act as if our house were on fire and fill the streets to demand change. Greta also said that "if standing up against the climate and ecological breakdown and for humanity is against the rules, then the rules must be broken." Noting that "conventional approaches of voting, lobbying, petitions and protest have failed," the decentralized Extinction Rebellion and others are engaging in civil disobedience in dozens of countries around the world. And in the first few months of Fire Drill Fridays, more than six hundred people joined Jane in risking arrest in Washington, D.C.

This all gives me great hope.

Civil disobedience can create the context for bold climate leadership in ways both pragmatic and powerful. As a tactic, it can shift political discourse and social context, creating the conditions for change. And on a personal level, it can be rewarding, empowering, and, many say, even transformational.

Many of us wrestle with the incongruence of our despair and anxiety about the climate crisis amid the demands of our day-to-day lives. We want to do more; we know we need to do more. Breaking out of our routine to engage in civil disobedience addresses that incongruence. In doing so, we join the long tradition of those who put their bodies on the line because the situation was that urgent. Aligning our values and our bodies feels truly liberating. When we started Fire Drill Fridays in Washington, D.C., many of the participants at our civil disobedience training said this was their first time doing so. This was less true toward the end, because many people who participated were so moved by the experience that they came back to do it again!

Civil disobedience can also create change in those who observe it. When Jane, and the students, nurses, workers, scientists, and others who joined her, engaged in civil disobedience, they inspired others to reflect on whether they, too, are moved to action. One person's inspirational action can move another person to act. That is part of the power of civil disobedience: to shift the context in which issues are considered, creating pressure to force leaders to act—which is exactly why we do it.

As Martin Luther King Jr. wrote in his Letter from Birmingham Jail,

> **Nonviolent direct action seeks to create such a crisis and foster such a tension that a community which has constantly refused to negotiate is forced to confront the issue. It seeks so to dramatize the issue that it can no longer be ignored. . . . The purpose of our direct-action program is to create a situation so crisis packed that it will inevitably open the door to negotiation.**

While there are many forms that civil disobedience can take, there are also clear guidelines that Greenpeace and others adhere to; above all, we commit to nonviolence, a principle that we never abandon.

Of course, not everyone chooses to, or can, participate in civil disobedience. There are well-documented patterns of institutional racism, discrimination, and abuse within the immigration and criminal justice systems in the United States. For very good reasons, people of color, immigrants, and others may choose not to get anywhere near police and jails. Others have health concerns, or family or work responsibilities they can't take a break from.

No problem; there are many ways to help advance climate solutions, and it certainly isn't necessary for every single person to engage in civil disobedience. But for those who can, I encourage you to learn more and connect with a group that can provide training and support for civil disobedience. Many cities have legal groups like the National Lawyers Guild that can offer legal advice and support. After three decades of climate activism resulting in no serious political or business leadership on climate at all, it's clear we need mass disruption to force action. Civil disobedience has worked before, and it can work again. Let's go!

Greenpeace
Nonviolence Guidelines

No matter the circumstance of provocation, we will not threaten others with physical harm, nor respond with physical violence to acts directed against us.

We will not call names, make hostile remarks, or otherwise inflict verbal violence upon others.

We will carry no weapons. If a necessary tool could be used as a weapon, we will never use it against a person or threaten a person with it.

We will minimize or eliminate any pollution or other harm to the environment that our activities could cause.

Our attitude—as conveyed through words, symbols, and actions—will be one of openness, politeness, creativity, commitment, and respect.

We will strive to speak to the best in all people, rather than seeking to exploit their weaknesses to what we may believe is our advantage.

We will always attempt to interpret as clearly as possible to anyone with whom we are in contact—and especially to those who may oppose us—the purpose and meaning of our actions.

We will adhere as closely as we are able to the letter and spirit of truth in our spoken and written statements.

We will take responsibility for and accept the legal and other consequences of our actions and will not seek to evade these consequences beyond legitimate recourse.

We agree to follow the directions of the decision-making body. In the event of serious disagreement where compromise cannot be reached, participants agree to remove themselves from the activity.

We will not damage property, except as necessary to remove barriers to our rights to free speech and assembly.

We will not bring or use any intoxicating drugs or alcohol, other than for medicinal purposes.

We will not run.

JANE FONDA is an Emmy award–winning actress, a two-time Oscar winner, and a political activist. She sits on the boards of V-Day: Until the Violence Stops, the Women's Media Center (which she cofounded in 2004), the Georgia Campaign for Adolescent Power & Potential, and Homeboy Industries. She lives in Los Angeles.